Cancer Simply Explained

Visar Vela • Besmira Sabani • Günther Spahn

Cancer Simply Explained

What is Cancer and What Can We Do About It?

 Springer

Visar Vela
Winterthur, Switzerland

Günther Spahn
Trimed Mainz
Karlsruhe, Germany

Besmira Sabani
Pharmaceutical Technology & Pharmacology
Zurich University of Applied Sciences
Wädenswil, Switzerland

ISBN 978-3-031-84296-2   ISBN 978-3-031-84297-9   (eBook)
https://doi.org/10.1007/978-3-031-84297-9

English translation of the original German edition published by Springer Nature

This Springer imprint is published by the registered company Springer Nature Switzerland AG
The registered company address is: Gewerbestrasse 11, 6330 Cham, Switzerland

If disposing of this product, please recycle the paper.

This book is for anyone interested in learning more about the disease in a simple and easy-to-understand manner without losing scientific and medical accuracy. Furthermore, it includes analogies to the real world that we haven't seen before in this way in any other cancer book. We will learn from the inner workings of our cells up to the different mechanisms of how cancer develops due to different internal and external factors. This book will also tackle how cancer is diagnosed and the possible treatments that are available to the patients. More importantly, it presents different practical ways of how cancer may be prevented and how we can strengthen our bodies to fight against this malady. Furthermore, there is much guidance and advice on what can be done about cancer. With that in mind, we hope you enjoy and benefit from this book as much as we do!

Preface

When confronted with a cancer diagnosis, one is often overwhelmed with apprehension and inquiries. This book is intended for individuals grappling with questions, including those puzzled by the singular development of cancer in multiple areas rather than distinct cancers. It addresses those pondering the paradox of developing cancer despite meticulous lifestyle choices and explores inquiries surrounding the pace of cancer progression in older individuals and its potential correlation with stress. Additionally, it seeks to clarify perplexities such as why postoperative chemotherapy is recommended despite complete cancer removal. Through enhanced understanding, we aspire to empower patients in self-care. This malady is an integral part of life for many of us and our loved ones.

Cancer is a disease that becomes more and more widespread as time goes by. Even though cancer has become more prevalent due to several factors, this disease has been around for a long time. The oldest documented case was from Egypt around 1600 BC, describing a breast tumor. This disease has been a part of life for many, including people close to us. An increasing phenomenon of people living in fear of developing the disease has become more prominent in today's society. Unhealthy fear has caused us to avoid regular checkups, as we dread being diagnosed with a disease that seems incurable. However, this choice may lead to suffering and early death.

We are lucky to live in a time when medicine has progressed, making cancer more treatable. For this reason, cancer mortality has been declining for years despite increasing diagnoses. Advancements in information dissemination have caused an increased awareness of the disease. It has become possible to produce helpful and medically accurate advice that is accessible to a broader audience. In addition, free research journals that discuss different aspects of cancer are published worldwide. However, these might seem difficult to digest for readers who only have a little scientific or medical background. That is why we created this book so that we could present concepts about cancer in a simple, easy-to-understand, yet medically and scientifically accurate way. We have created many analogies that will help you understand cancer better. Here, we discuss cancer from conception to diagnosis, treatment, and finally, therapy. This book also presents various ways to inform yourself about cancer and prevent disease development, such as through proper nutrition and physical activity. Remember, these concepts and information are for education only; in no way does this book replace a visit to a doctor.

Winterthur, Switzerland   Visar Vela
Wädenswil, Switzerland   Besmira Sabani
Karlsruhe, Germany   Günther Spahn

Competing Interests The authors have no competing interests to declare that are relevant to the content of this manuscript.

Preliminary Remarks

This book is for anyone interested in learning more about cancer in a simple and easy-to-understand manner with scientific and medical accuracy. We will look at a wide range of topics, from the inner workings of our cells to the different mechanisms of how cancer develops due to various internal and external factors. In addition, we will discuss how cancer is diagnosed and what possible treatments are available today.

More importantly, in this book, we present different practical ways in which cancer may be prevented and how we can strengthen our bodies to fight against this disease.

There is plenty of guidance and advice on what can be done about cancer. There are many tips on what to ask your doctor during diagnosis and treatment.

Finally, you will find practical tips in the chapters on nutrition and physical activity to prevent cancer. This book highlights the significance of maintaining healthy nutrition and regular exercise to optimize the body's cancer-fighting mechanisms. With that in mind, we hope you enjoy reading this book and benefit from it!

Acknowledgments

During this journey of writing about cancer, we met some wonderful people who helped us to realize this book. First of all, we would like to thank our *families* and *friends* for their unwavering love and assistance throughout this journey. A special thanks goes to *Dr. med. Laurence Favet* who provided valuable comments and suggestions in the treatment section to improve this book. We would like to thank Professor Hamilton *Craig,* Ideja *Bajra and Vegim Kamberi* for their academic editing and critical proofreading. Furthermore we would like to thank Dr. Erand *Llanaj* for the inputs in the section nutrition. Finally, we want to thank everyone at Springer, particularly *Stafanie Wolf* and *Ina Karen Stoeck*, for their unwavering support and assistance in the realization of this endeavor.

Declaration of Generative AI

During the preparation of this book, the authors used ChatGPT only to find relevant information for the newest cancer treatment studies. Afterwords, the authors reviewed and edited the content and take full responsibility for the content of the final publication. The authors also used AI programs to draw many of the figures in the book (DALL-E).

Introduction

Cancer is a disease that is slowly becoming the most common cause of death, especially in countries where junk food is abundant and physical activity is lacking. Cancer is not contagious, yet it is spreading rapidly. According to the World Health Organization (WHO), 20 million new cancer cases were reported in 2020, when 10 million people died of the disease [1]. These figures increase every day worldwide. According to the forecast, the number of people who die from cancer each year could increase to about 35 million by 2050 [2]. These numbers are one reason most of us have lost friends and family members to this disease or know someone who has been diagnosed with some form of cancer.

Cancer incidence worldwide is unevenly distributed. As Fig. 1 shows, the rate in highly industrialized Western countries is high, with reportedly around 450 cancer cases for every 100,000 inhabitants. Lung, colon, breast, and prostate cancers are widespread in those countries. Meanwhile, rates reported in Asian countries are lower, with only around 97 cases per 100,000 individuals. In those countries, stomach and liver cancers are more widespread than other cancer types [3].

Figure 1 also reveals that low rates of cancer are reported in Africa. However, cancer incidence and mortality have risen in Africa over the last few years. As cancer incidence increases with age, higher life expectancy on the continent partly explains this trend. Unfortunately, most of the population still lacks awareness about the disease and strategies to prevent it. Cancer research and health care have not been prioritized, and reports and statistics on cancer rates in Africa are often spread across different registries. There is also a shortage of medical equipment to detect and treat cancer, as well as medical staff and funds to combat the disease [4]. Yet, unlike Africa, Australia has a very high

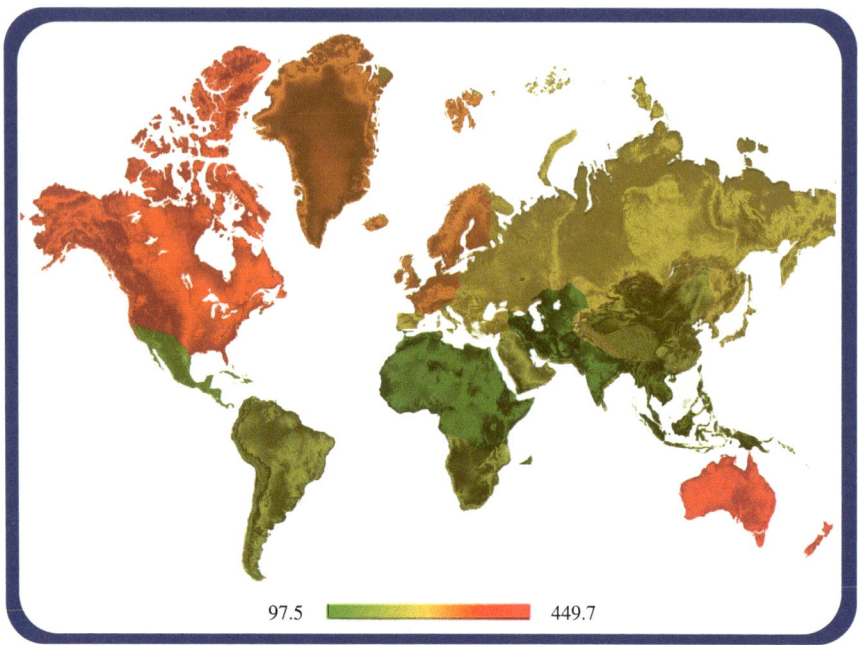

Fig. 1 Worldwide distribution of cancer incidence. The map shows the incidence of all cancers for every 100,000 inhabitants

rate of cancer, which is attributed to certain factors, such as geographical location, sun exposure, and migration. Interestingly enough, cancer mortality in wealthy countries is only about 15% lower than in developing countries [3]. These factors will be further discussed in Chap. 2 ("Cancer Development") in the section titled "Cancer Risk Factors." What is important to realize is that even though rates vary worldwide, cancer is nevertheless everywhere.

Because we are more likely to die from cancer than from a car accident [5, 6], cancer continues to interest many people. As Fig. 2 shows, the risk of developing cancer and dying from it is very high in general. Fortunately, continuous research and analysis over the years have taught us much about cancer, so our knowledge has continually expanded. This is illustrated in Fig. 3.

However, despite our knowledge, certain questions remain to be answered. Are we powerless against this disease, or is there something we can do? How can we protect ourselves from cancer and live an enjoyable life? This book will answer those questions and provide you with a wealth of knowledge regarding cancer.

Fig. 2 Fears vs. actual risk. Developing cancer has a very high risk, next to the risk of death caused by the effects of smoking [7, 8]

This book will also try to offer information simply, with the help of analogies, but without much complicated scientific jargon. Although you may have heard some of these terms before in your daily life or biology classes at school, some details might seem complicated. Do not worry if you do not understand all the concepts at first. Everything will slowly become clearer, chapter by chapter. Since cancer often seems very abstract, we have tried to use familiar images and comparisons to help you better visualize our ideas. We will discuss cancer as a disease, describe its mechanisms, development, diagnosis, and treatment, and explain how nutrition and physical activities help to prevent the disease and limit its progression. As the numbers reported by the WHO suggest, we must take action now. It is in everyone's interest to do so. But first, we need to understand some mechanisms in the human body to get a clearer picture of cancer.

10 INTERESTING FACTS ABOUT CANCER

"CANCER" COMES FROM THE GREEK WORD *"KARKINOS,"* WHICH MEANS CRAB.

Early physicians referred to tumors that formed branching veins or offshoots from the main body as crab-like or 'cancerous.'

MORE CASES OF SKIN CANCER ARE LINKED TO TANNING BEDS THAN LUNG CANCER IS TO SMOKING.

In the U.S. alone, more than 419,000 new cases of skin cancer are caused by tanning beds each year. However, skin cancers are usually detected at an earlier stage than lung cancer.

EGYPTIANS WERE THE FIRST TO DESCRIBE CANCER

The Edwin Smith Papyrus describes breast tumors removed by a 'fire drill' tool but mentions no treatment. This document is dated back to 1600 B.C.

OVER 50% OF ALL CANCER CASES CAN BE PREVENTED

Researchers believe that about 2.4 million cancer cases and 3.7 million cancer deaths are preventable annually

MORE THAN 200 DIFFERENT TYPES AND SUBTYPES OF CANCER EXIST

Over 200 various types and subtypes of cancer have been identified. New cancer subtypes are discovered daily

AROUND THE WORLD, 28 MILLION PEOPLE SURVIVED CANCER

Although cancer cases are rising, more people survive cancer than ever before, thanks to the advances made in recent decades

NAKED MOLE RATS DO NOT DEVELOP CANCER

The tissues of these African mammals are rich in a certain kind of hyaluronic acid, which inhibits cancer growth

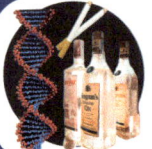

ONLY 5 TO 10 PERCENT OF ALL CANCERS ARE HEREDITARY

The majority of cancers are caused by a mix of inherited and environmental factors, such as diet, alcohol, smoking, and obesity

BREAST CANCER IS MORE LIKELY TO DEVELOP IN THE LEFT BREAST THAN IN THE RIGHT

Cancer in the left breast is 5%-10% more frequent than cancer in the right breast

ELEPHANTS HAVE INCREDIBLY LOW CANCER RATES

They possess additional copies of two genes that fight cancer: *P53*, which looks for cells with DNA damage, and *LIF6*, which gets rid of abnormal cells before a tumor forms

Fig. 3 Ten interesting facts about cancer. Cancer remains a very elusive disease, with new information being discovered every second [9–11]

References

1. Cancer Today. https://gco.iarc.who.int/today/. Accessed 04 July 2024.
2. International Agency for Research on Cancer. Latest global cancer data: cancer burden rises to 19.3 million new cases and 10.0 million cancer deaths in 2020. 2020.https://iarc.who.int/wp-content/uploads/2020/12/pr292_E.pdf. Accessed 06 July 2024.
3. Sung, H. et al. "Global Cancer Statistics 2020: GLOBOCAN Estimates of Incidence and Mortality Worldwide for 36 Cancers in 185 Countries." CA Cancer J Clin, 2021, 71(3), 209–249. https://doi.org/10.3322/caac.21660.
4. Hamdi Y, et al. Cancer in Africa: the untold story. Front Oncol. 2021;11:650117. https://doi.org/10.3389/fonc.2021.650117.
5. Cancer. https://www.who.int/news-room/fact-sheets/detail/cancer. Accessed 06 July 2024.
6. Shapiro A. The book of odds. HarperCollins; 2013.
7. An update on cancer deaths in the United States.
8. CDC. Heart disease facts. Heart Disease. https://www.cdc.gov/heart-disease/data-research/facts-stats/index.html. Accessed 06 July 2024.
9. Roche | Interesting things you may not know about cancer. https://www.roche.com/stories/9-things-about-cancer. Accessed 06 July 2024.
10. 8 Early signs of breast cancer: what to look for? | The University of Kansas Cancer Center. https://www.kucancercenter.org/news-room/blog/2020/07/8-early-signs-of-breast-cancer. Accessed 06 July 2024.
11. Elephants rarely get cancer thanks to a 'zombie' gene. https://www.science.org/content/article/elephants-rarely-get-cancer-thanks-zombie-gene. Accessed 06 July 2024.

Contents

About the Authors

Visar Vela Is a medical diagnostic assistant. His passion for cancer research and diagnostics arose while working on his master's thesis, which involved researching blood cancer in Adrian Ochsenbein's lab at the University of Bern. During his PhD in biological medicine, he uncovered the mutational landscape of a rare lymphoma type at the Department of Pathology in Basel under Stefan Dirnhofer and Alexandar Tzankov. By writing this book, he hopes that the reader will get the best possible information about cancer development and cancer diagnostics.

Besides having a passion for cancer research, Dr. Vela loves jogging, sailing, and cycling. He grew up in the lovely Swiss Alps, in Brig. He still has his most creative phases at the airport. He says that airports are great sources of problems that need solving, such as cancer research.

Correspondence: Visar.Vela@gmail.com

Besmira Sabani A dedicated PhD candidate at Zürich University of Applied Sciences in Wädenswil, Zurich, delves into the innovative field of pharmaceutical technology and drug discovery in Steffi Lehmann's lab. With a strong focus on the development of potential drug delivery systems, Sabani's work highlights her journey from her early academic years to her first publication on novel functionalization strategy for the target delivery of extracellular vesicles (EVs) to cancer therapy. She is passionate about finding new approaches in nanotechnology enabling the targeted delivery of drugs to their site of action. With this book she hopes that the reader will get more information about the latest cancer treatment technologies.

Besides having a passion for cancer research, Sabani loves cycling and solving puzzles, similar to cancer, whose complexity is puzzling us for decades. She grew up in Minusio, Ticino.

Günther Spahn Studied medicine in Tübingen. For his MD in immunology and vaccine research, he worked at the Luxembourg National Institute as a pioneer in the field of measles vaccine research with Prof. Claude Muller. For his biomedical postdoc, he joined Prof. Pecher at Max Delbrück-Institute in Berlin in the field of cancer immunology before starting his clinical career as an academic trainee at Charité Berlin in hematology and oncology. As a scientific member of the leukemia and lymphoma network in Germany, he worked as a consultant in multiple clinical centers. Today, he has his own outpatient clinic in Mainz. His main focus is supporting patients with cancer and their families in all aspects of care: prevention, treatment, rehabilitation, complementary medicine, and palliative care in the field of cancer medicine. With this book, he wants to teach us how to fight cancer before it starts and the options we have besides conventional treatment.

1

Cell and Cancer

Contents

Abstract This chapter explores the intricacies of cells and their fundamental role in both health and disease, with a particular emphasis on cancer. With approximately 37 trillion cells cooperating to form tissues and organs, cells—the smallest unit of life—are absolutely necessary for the proper functioning of the human body. Every cell, from the nucleus to the mitochondria, contributes to vital activities like energy and communication. The nucleus stores DNA, which contains genetic instructions and is essential for regulating cellular activity and features. When mutations develop in DNA, it can result in malfunction of the cell cycle, which is responsible for controlling division and renewal, ending in uncontrolled growth and cancer. This chapter emphasizes how disturbances in cellular communication and energy balance, often associated with mitochondria and plasma membranes, can trigger tumors. When it comes to understanding the complexities of cell structure and function, it is essential to have a firm grasp on how these minuscule units are responsible for making life possible and how mistakes can result in diseases including cancer.

1.1 The Cell

A cell is the smallest unit of life, yet it is wonderfully complex. Cells are so small that they measure only around *10–100 micrometers* in diameter. Thus, they cannot be seen with the naked eye. Hypothetically, we could fill a single water droplet with cells equal in number to the population of Switzerland's two biggest cities, Zurich and Geneva. Once a cell becomes abnormal, it starts to multiply uncontrollably. For this reason, we'll explain what a cell is. The origin of these abnormalities lies in its genes, its DNA. In this chapter, we'll explain what DNA is and what genes correspond to.

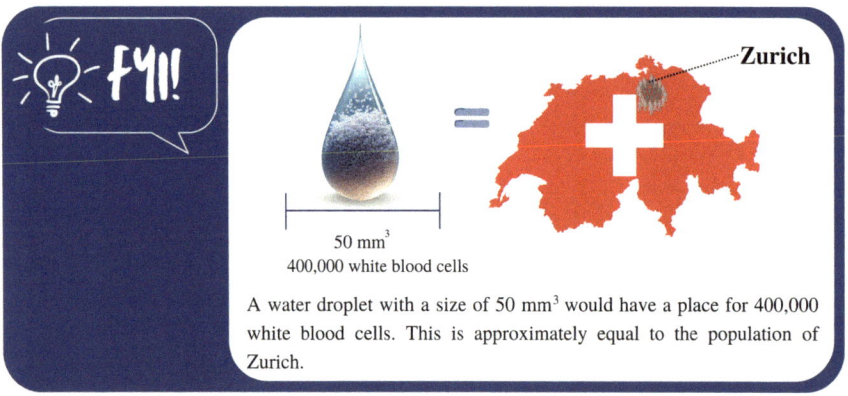

50 mm^3
400,000 white blood cells

Zurich

A water droplet with a size of 50 mm^3 would have a place for 400,000 white blood cells. This is approximately equal to the population of Zurich.

The human body contains numerous cells that are difficult to quantify. So far, the best estimate is that the human body has more than *37 trillion cells* [1]. If you place all 37 trillion cells side by side, you will form a chain four million kilometers long. A chain of all the cells in a single human body could thus *circle the Earth about 100 times* (Fig. 1.1). However, this only refers to the number of cells in a body at a certain time. Many of these cells become useless and are constantly replaced by new ones. Of those *37 trillion* cells, around *50 million die every second,* but they are immediately replaced by their exact copies.

The entire human body is made up of various types of cells. Cells are like the tiny living bricks that make up the human body. They are very small but wonderfully complex. Like certain fruits, for example, they have an envelope and a nucleus, in which, as we'll see later, their DNA is stored, with the

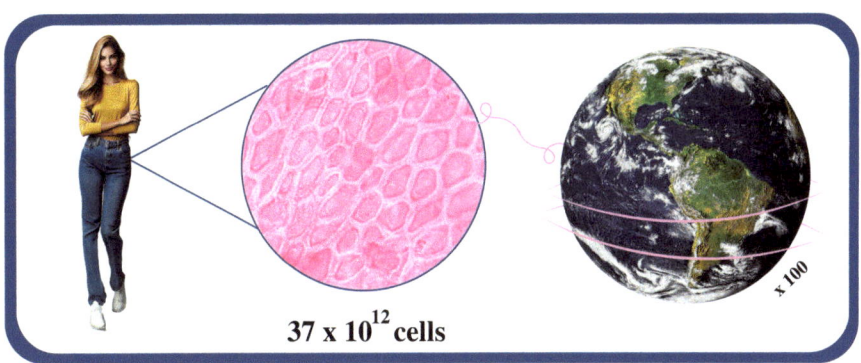

Fig. 1.1 The extent of human cells. When the 37 trillion cells in a human body are placed side by side, they will form a four-million-kilometer-long chain that circles the Earth around 100 times

exception of red blood cells, which have no nucleus. Together, they create a living being that can perceive and interact with its environment. While the estimated number of cells in a human body is 37 trillion, there are roughly 300 billion muscle cells, which contribute the most to body mass. Approximately 30% of a person's total body weight can be attributed to muscle cells. In comparison, an average of 39 trillion bacteria represents just 0.3% (100–200 g) of a person's total body weight [1, 2]. In addition, 30 trillion of the 37 trillion cells are red blood cells. They thus constitute the majority of cells in the human body but are only 3.6% of a person's total body weight.

In his book *Immune: A Journey into the Mysterious System That Keeps You Alive* (2021), Philipp Dettmer compares the interaction of cells to the teamwork of ants. On its own, an ant is completely helpless. However, it can accomplish wonderful things if it collaborates with other ants in the colony. Each ant has abilities that may be unique. The ants exchange information, interact with each other as a team, and build complex structures, all with the goal of survival (Fig. 1.2a).

Similarly, each cell is nothing more than a bag full of small cellular structures controlled biochemically. On its own, a cell cannot do anything. Nevertheless, if cells work together, they can accomplish incredible things. (Fig. 1.2b) They can build specialized tissues and entire organ systems. They range from muscles that create your heartbeat to brain cells that allow you to think and understand this book in your hands. Biochemical reactions

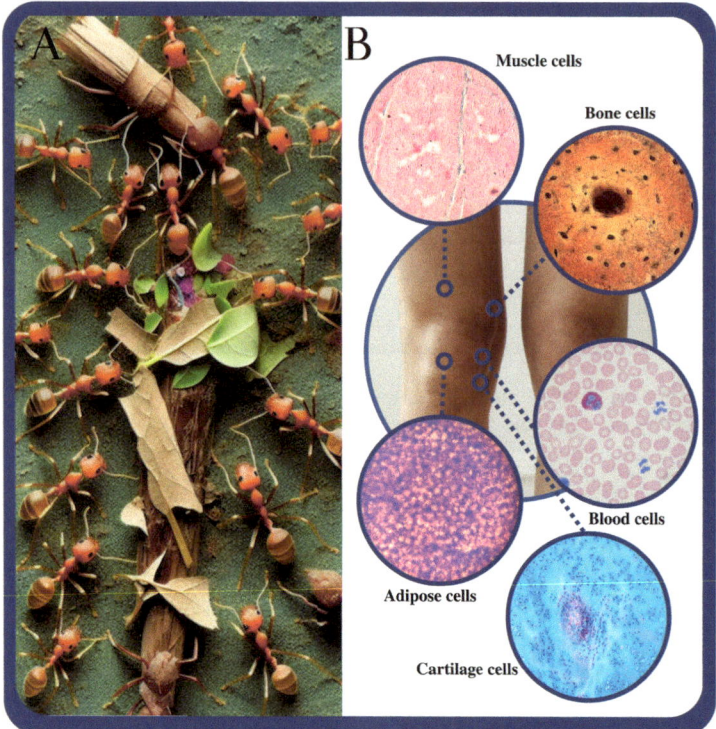

Fig. 1.2 Interaction is needed for survival. (**a**) Ants cooperate to accomplish a common goal. (**b**) Similarly, different cells perform different roles but work together to perform a specific activity

power a cell entirely; it absorbs nutrients, grows, excretes waste, responds to stimuli, and can reproduce. These processes are directed by even smaller parts of cells.

A cell works like a factory, where essential functions are distributed to different departments to meet optimal conditions for the efficient manufacturing of products (Fig. 1.3).

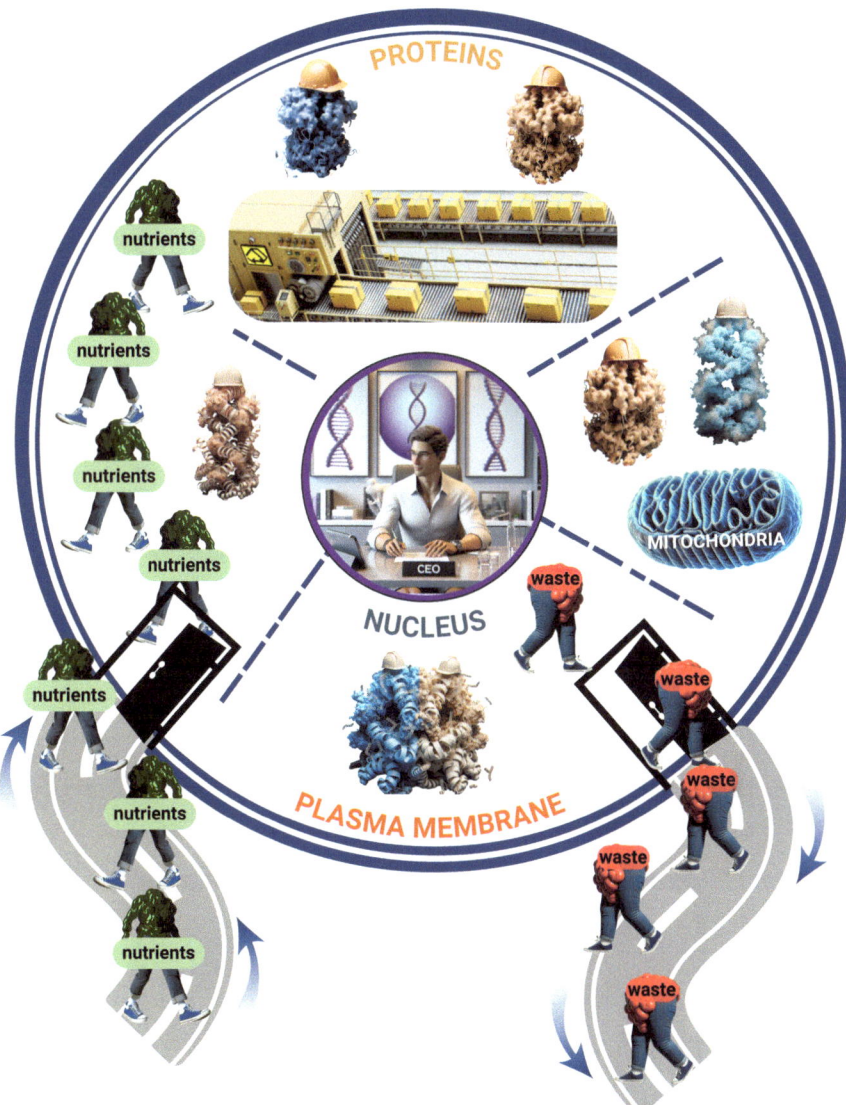

Fig. 1.3 A cell can be compared to a factory. Different departments perform different essential tasks to accomplish the factory's intended purpose

NUCLEUS

The *nucleus* is like the factory's Chief Executive Officer, or *CEO,* who controls all the factory's activities. The *nucleus* determines what proteins need to be made to properly function in response to environmental changes, such as inducing cell division. The nucleus is the cell's information center, which houses the DNA, where genetic information is encoded. The cell's information center is like a big library, containing about 750 books, equivalent to 1.5 GB of data or roughly 300 songs.

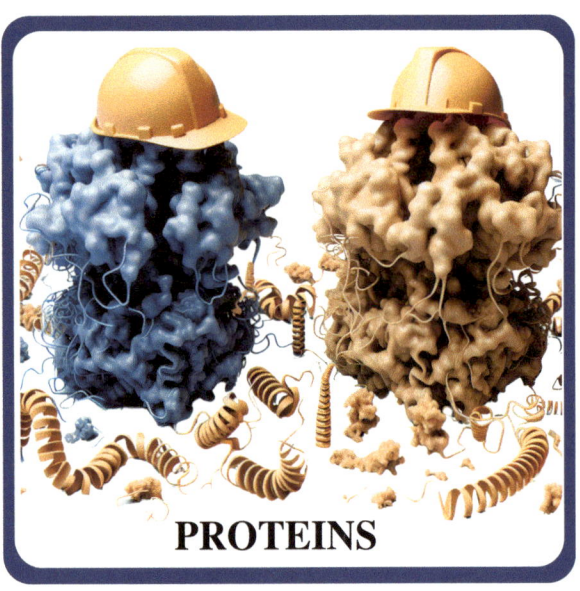

PROTEINS

The **proteins** are the factory's workforce; they carry out the functions necessary to maintain cells and keep them in their optimal state to fulfill their designated roles. These proteins transport nutrients from the blood, communicate with the outside to identify external changes, and transform nutrients into energy.

Inside the cell, there is a specialized transport network, a packaging center, parts for digestion and recycling, and also building centers. Proteins can form **enzymes** that act as engineers and transform unusable materials into essential substances, enabling cells to adapt to environmental changes. Enzymes also ease communication between cells by relaying signals and creating circuits that coordinate with the cells' life cycles. While adaptation is vital for survival, when cells fail to adapt to different environmental changes, this usually causes enzyme and protein function errors, which, in turn, can allow cancer to develop [3].

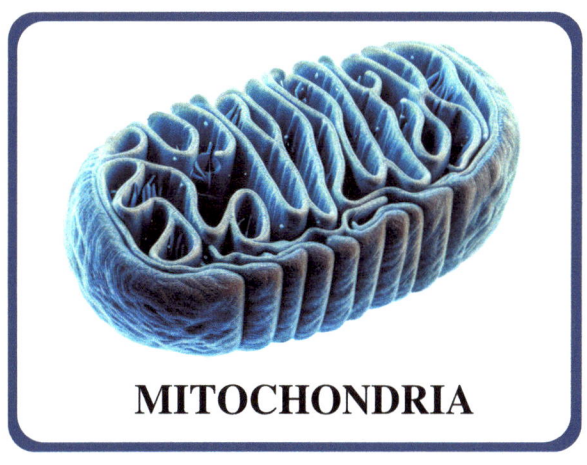

MITOCHONDRIA

Meanwhile, the **mitochondria** serve as the cell's power source, where energy stored in molecules (sugar, proteins, and fats) is converted into cellular energy. Mitochondria are microorganisms that have a symbiotic interaction with human cells. They have their own DNA and cell membranes and can reproduce outside the cell cycle. Cell division randomly distributes them among the daughter cells. Multicellular organisms, such as humans, develop a distinct hereditary line due to their own genetic material [4]. The mitochondria in an egg cell serve as the starting point for all subsequent mitochondria in an organism. Damage to these mitochondria can lead to catastrophic disorders. Because males do not pass on mitochondria, only maternal mitochondria permanently change their children's metabolisms. **Oxygen** fuels this conversion process, but it also produces toxic waste products called reactive oxygen species, or "free radicals." These free radicals introduce genetic changes called **mutations,** which also cause errors in making proteins for the body. The term "mutation" comes from the Latin word *Mutare,* which means *to change.* Meanwhile, free radicals are toxins, and the ability of our bodies to deal with them varies from individual to individual, is genetically determined, and, most importantly, influences cancer risk.

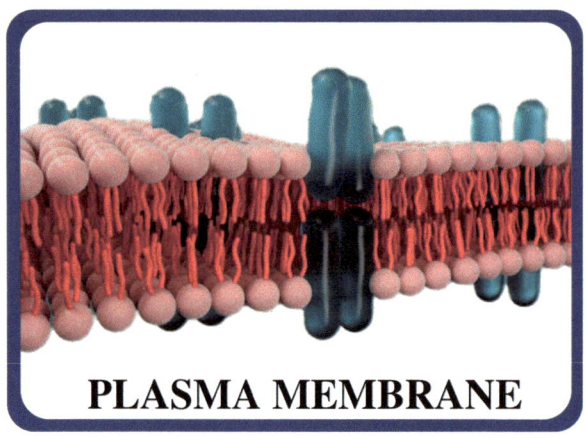

PLASMA MEMBRANE

Finally, the **plasma membrane** can be compared to the shipping and receiving department of a factory. It acts as a wall that contains all activities of the cell and forms a barrier between the cell and the external environment. The plasma membrane filters substances that enter and leave the cell. Some proteins in the plasma membrane detect chemical signals and transmit them to the nucleus. This kind of communication is critical since incorrectly reading and transmitting information could have disastrous results. When the cell no longer understands what is happening in its environment, it begins to behave independently. This erratic behavior also affects nearby cells and can lead to cancer.

In short, these proteins are just a few examples of the tens of millions of molecules in a single human cell. While roughly half of the molecules are viscous **water molecules** that allow other molecules to move around smoothly, a cell also contains other cellular components, each with its own specific function.

1.2   The DNA

We can imagine our DNA as a recipe that carries all our information and a set of genetic instructions that provide templates for proteins to build the human body (Fig. 1.4). Our DNA says what we will look like and how we will function. For instance, our DNA determines the color of our eyes or how our lungs work. [5, 6] Furthermore, DNA can also determine our physical fitness, intelligence, and health.

Ironically, only four letters are used to write all the vast information in our DNA. The letters A, G, C, and T are called ***DNA bases,*** and they stand for adenine, guanine, cytosine, and thymine. These DNA bases come in complementary pairs: A and T and C and G (Fig. 1.5). This pairing is essential to maintaining cell stability by providing a template for repairing the damage caused by any mistakes during DNA replication.

Each cell carries about 3.2 billion base pairs, known in their entirety as the ***human genome.*** The human genome is like an extensive library with many cookbooks containing thousands of recipes. Some recipes have information that is very useful to us for dishes we might often prepare. Therefore, we read them carefully. On the other hand, we also have recipes that are of little interest to us, but they remain in place in case someone is interested. This is similar to what happens in our genome: not all the information within the long line of 3.2 billion letters is useful. It can happen that genes are "locked" in certain cells, which is beneficial when you want a gene to remain silent, essentially preventing a "recipe" from being read. On the other hand, if the genes responsible for the cell's proper functioning are "locked" this can lead to cancer. One of the best-known ways in which the body "locks" DNA is called "methylation." Methyls are a kind of lock attached to certain parts of DNA, which prevents the reading of certain genes without modifying the genes themselves (Fig. 1.6). When the reading and expression of genes are prevented without altering the genes themselves, we speak in biological terms of "epigenetic" modifications, a word you've probably heard before. The gene remains intact but can no longer be read. The methylation (i.e., locking) of certain genes (e.g., protective genes) can thus contribute to the development of cancer. On the other hand, there is currently no cancer treatment that can act directly on methylation, but this is a subject of scientific research. Will it be possible in the future to "relock" proliferation genes or "unlock"' protective genes?

However, given all this information, we might wonder how similar we are with all these base pairs in our genome. While someone might say there is a wide variety between two humans, we are ***99.9% identical*** in our genetic

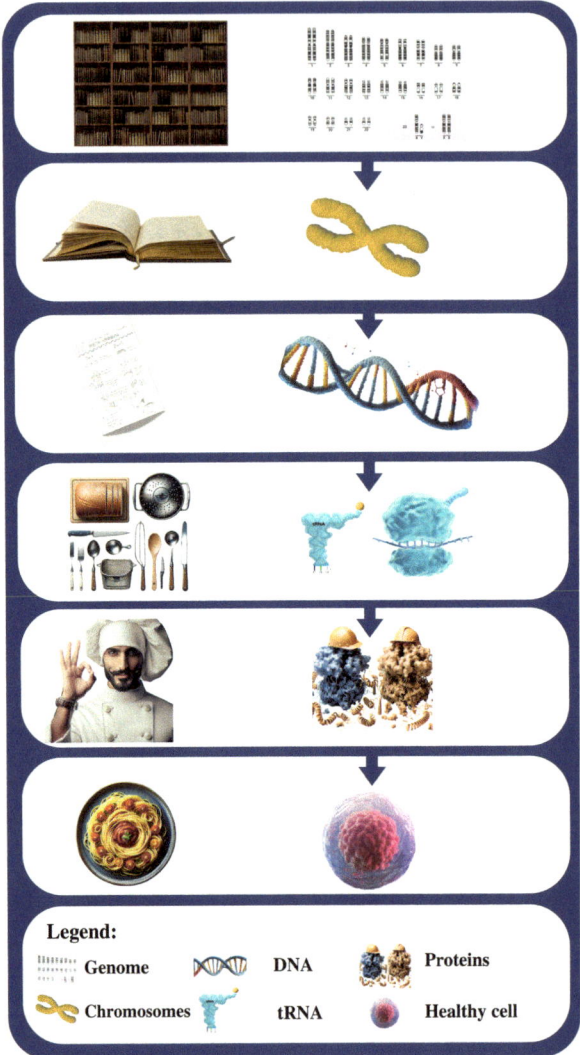

Fig. 1.4 DNA carries all our biological information. The human genome is like a library that contains numerous cookbooks that provide recipes to cook specific dishes using the right kitchenware. Similarly, the DNA in our chromosomes serves as a template for determining our characteristics and functions. It is a code with individual sections that represent all the instructions needed to build our cells

makeup. Just 0.1% of the total that is different makes each of us unique (Fig. 1.7). In other words, we are all more similar than different. Over 20 years ago, scientists needed many years to decode the sequence of the human genome, as the project's computational costs and data storage requirements

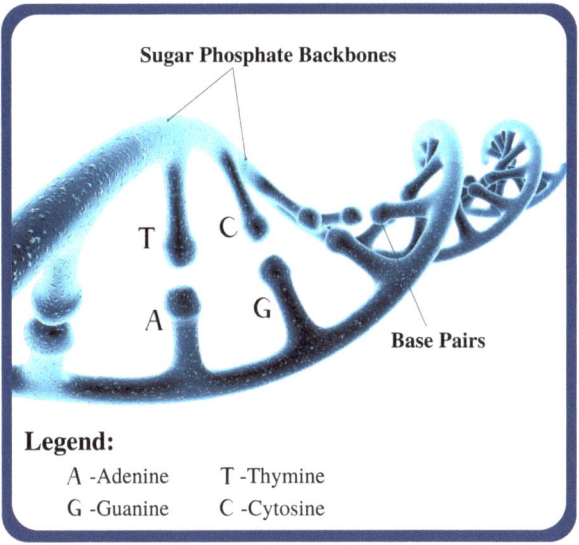

Fig. 1.5 DNA is a double-stranded molecule. It consists of bases that connect to one another to create a twisted ladder, also called a double helix

Fig. 1.6 Methyl groups act like locks that attach to certain parts of the DNA, blocking the reading process without altering the genetic code itself

Fig. 1.7 Humans are 99.9% genetically identical, but the remaining 0.1% varies significantly among individuals. This small percentage accounts for our unique individuality, determining our distinct appearances and subtle differences from our parents or children. Remember, no two humans share an exact copy of their DNA

were huge at the time. Yet, obtaining the sequence of the human genome told us the order and number of base pairs carried by different sections of our DNA. This provides us with specific information in sections called *genes.* Each gene exists in duplicate. We receive two alleles of a gene, one from our mother and one from our father, due to this duplication. Our genes are made up of DNA, which is organized in long strands and wound around proteins. In turn, DNA makes fibers that are organized into big molecules known as *chromosomes* (Fig. 1.8). Each chromosome is composed of two "reflecting" strands of DNA. Humans have *46 chromosomes,* divided into *23 pairs,* one set from each parent.

While our genes are arranged in the same set of chromosomes, what do they do? Genes play many important roles in the human body. Aside from defining our physical characteristics, such as our height or eye color and our susceptibility to diseases, genes also tell our cells to make proteins and perform various tasks. For example, our DNA does not tell our body to "make curls." Instead, our DNA tells our cells to build curl proteins, which, in turn, can make curls. This is how our different traits are created, thanks to the 0.1% mentioned before [7].

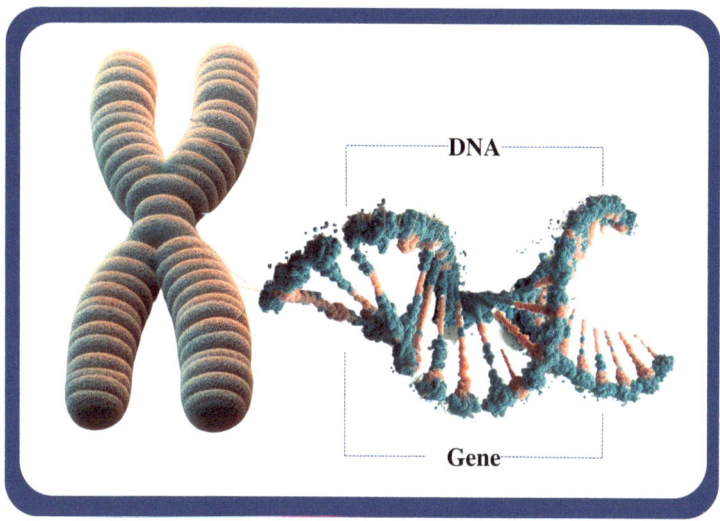

Fig. 1.8 DNA in chromosomes. Long DNA strands are wound around protein structures called chromosomes

During the Human Genome Project 20 years ago, scientists discovered we have just **25,000** genes, which is only 3% of our genome. Therefore, there is text in the "chromosome books" that is not grouped into "gene chapters." The full usefulness of the texts that are not organized into "gene chapters" is not yet fully understood and is the subject of ongoing research. Each cell will "use" a high proportion of a certain number of genes and "ignore" others, depending on its role in the organism. For instance, a pancreas cell will utilize the gene (open and read the chapter) responsible for manufacturing insulin, unlike a skin cell, which generally keeps this "gene-chapter" closed. Similar to the pancreatic cell, the skin cell has this "chapter-gene" in its "library" but does not refer to it. This gene must remain "locked" within this cell, acting as a "forbidden chapter." Contrary to healthy cells, cancer cells are sometimes capable of opening forbidden "chapter-genes" that should remain closed. For instance, a bronchial cell in the lung ordinarily should not be able to "open" the "chapter-genes," allowing it to move outside the lung. However, a cancer cell might be able to activate the "gene-chapters" that permit it to depart from its original location and migrate to other organs.

If we recall our cookbooks, even though we have many recipes, not all of them interest us. We may only have two or three recipes we like to prepare and which make sense to us, while the others may be less interesting. Before the Human Genome Project was completed in 2003, scientists first estimated our genome to have between 30,000 and 100,000 genes. Nobody expected the final number to be just **25,000 genes.** It is a number similar to that of a single-cell parasite.

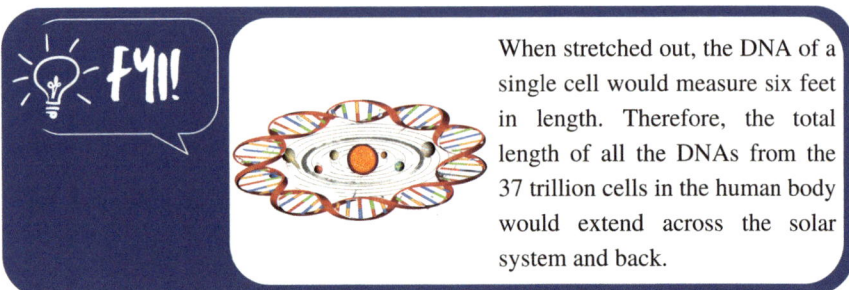

When stretched out, the DNA of a single cell would measure six feet in length. Therefore, the total length of all the DNAs from the 37 trillion cells in the human body would extend across the solar system and back.

But does size really matter? To find out, some scientists explored the genome of the marbled lungfish. Its genome has 132 billion base pairs, with many more genes than we have. In addition, its individual cells have 40 times more DNA than human cells do. However, the biology of a marbled lungfish is less complex than ours [8]. This is due to its enormous genome, which is filled with repetitive and non-coding DNA, such as transposable elements, illustrating the C-value paradox—more DNA does not necessarily mean greater complexity. While the lungfish has an abundance of genetic material, much of it serves non-functional or regulatory roles, whereas human complexity arises from intricate gene regulation networks and specialized cellular functions.

1.3   The Cell Cycle

Cell division is a naturally occurring process to replenish cells that die. Every second, more than **500,000 cells** die in a typical adult body [9]. One reason is that some of our body cells deplete themselves faster than others. For instance, cells at the skin surface and inner walls of the intestines are exposed to constant mechanical stress. Therefore, they are replaced after a few days. Cells that are constantly exposed to these stresses are particularly vulnerable and more likely to die. Therefore, they need to be replaced more frequently to ensure a perfect seal between the outside world and the inside of the body. Similarly, the body's organs, which often change under the influence of hormones, have phases of cell proliferation and destruction. Each time the cell divides, it must "photocopy" its DNA, and this is when "photocopy errors" can occur. Important "chapter genes" can be lost, and "forbidden chapters" can open up and be read, leading to cancer. This is also the case with the cell layers of our airways; due to their constant exposure to toxic substances, their DNA is more prone to making mutations. In fact, constant exposure to toxic substances is why cells in our intestines, stomach, and respiratory tract have to renew themselves regularly. In contrast, cells that divide very little rarely cause cancer. Neurons, the nerve cells, have a very long-life expectancy (Fig. 1.8), and the production of new neurons is very rare. As a result, there is very little risk of developing DNA errors, and these cells rarely cause cancer.

Our blood cells are the champions of self-renewal. Within a few hours after they die, our red blood cells are replaced so they can continue to perform their important tasks. Because red blood cells lack nuclei, they cannot directly undergo mutations and consequently cause cancer. When red blood cells proliferate excessively, it is a disease originating from their stem cell, known as the ***mother cell***. As we mentioned, red blood cells originate from one precursor cell, known as a ***stem cell***. These stem cells can develop into any cell type, and they renew all types of blood cells. This makes them different from most other cells, which divide themselves more or less ***40 times*** before they die forever. In contrast, stem cells are immortal, placed in safe niches of our organs, which means they represent the cell reserve of our body [10]. Meanwhile, other cell types, such as our nerve cells, have longer life expectancy (Fig. 1.9). While cell death and cell renewal are a permanent process, it takes ***80–100 days*** for our body to replace all of its 37 trillion cells completely [11].

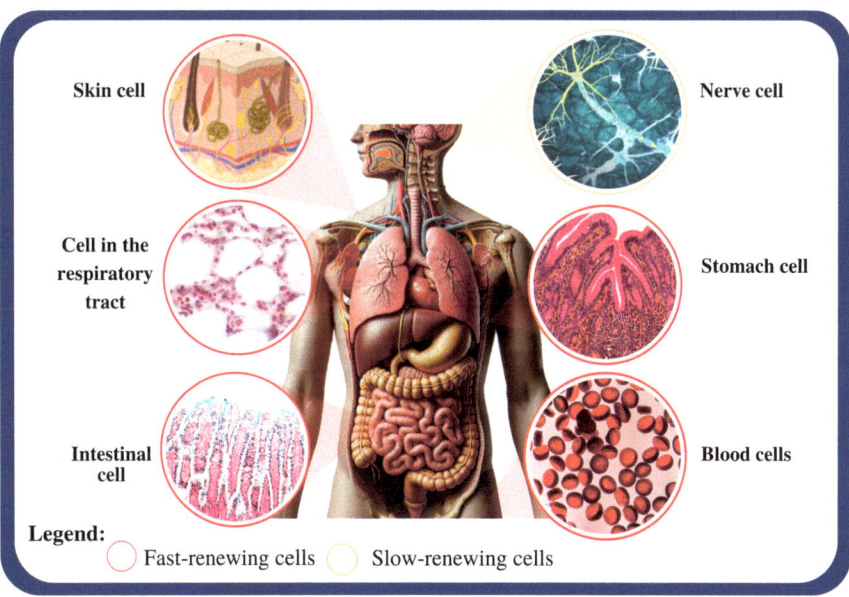

Fig. 1.9 Different types of cells renew at varying rates. Cells in the skin, digestive and respiratory tracts, and blood, due to constant exposure to mechanical stress and toxic substances, renew rapidly. Meanwhile, nerve cells in the brain, due to their complex nature, renew slowly

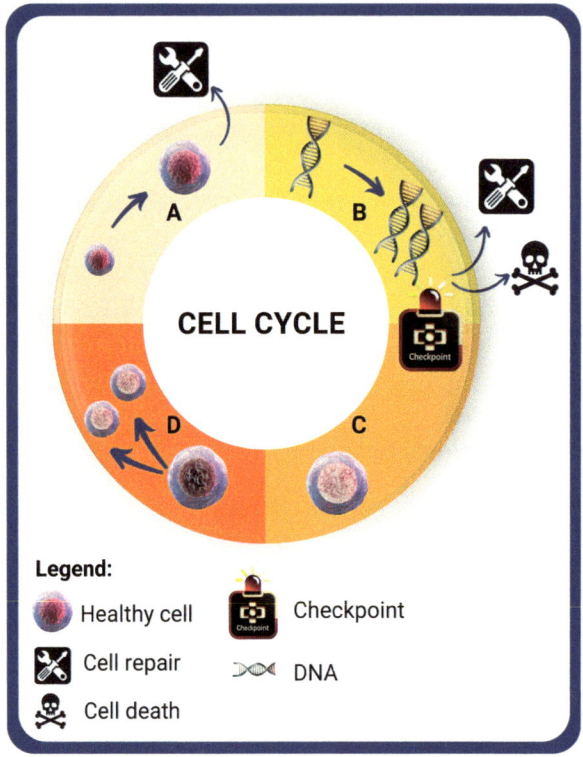

Fig. 1.10 Cell cycle. The different phases of the cell cycle and the specific checkpoints determine control, where the end goal is to have a functional cell division. Cell division is essential to keep the organism running

Cell renewal is a constant process that must run under controlled conditions. This is because we have ***several million cell divisions*** every second [12]. During this orchestra of work, all the genetic information equal to ***750 books*** (1.5 GB) has to be copied to form daughter cells. Due to the volume and speed of this process, there is a high probability that our new cells will acquire mistakes (mutations) even if we do not want them to. For this reason, the cell cycle ensures that cell division usually runs smoothly. The nucleus of the cell and other specialized structures must be duplicated to produce a fully functional daughter cell.

The cell cycle is the process by which a cell divides into two different cells. To achieve this, the cell needs to increase its size by doubling (Figs. 1.10a, b) until it has enough material for two. It also checks carefully to make sure the DNA is not damaged. If any damage is found, the cell cycle is stopped for repair. Then, the ***DNA is replicated, which means our DNA doubles*** [13, 14]. During copying, mistakes or mutations can occur, resulting in changes to the DNA sequence that can contribute to cancer development. As mentioned earlier, there is a higher risk of mistakes in cells in the intestines, stomach, and respiratory tract because they are more exposed to toxic substances and because they divide more frequently than other cells.

After the DNA is replicated, a check is made to ensure that everything has been completely doubled without damage. If the cell detects damage, it starts to repair itself. If the damage is beyond repair, the cell will initiate programmed cell death, known formally as *apoptosis,* to prevent the mistakes from being passed on to other cells. After completing this step, the cell prepares for cell division (Fig. 1.9c). Then, after successfully dividing, the cell cycle starts again [15]. (Fig. 1.9d) All of these processes generally take 24 h to complete [12].

You may have noticed in Fig. 1.9 that we have control mechanisms or checkpoints to prevent harmful mistakes from being passed on to other cells. These repair mechanisms are very accurate. Without repairs or programmed cell death (apoptosis), mistakes would accumulate in each new process of division and favor cancer development. Thanks to macrophages, the largest immune cells that engulf dead cells, the body can easily get rid of dead cells.

Most cells in your body have a limited lifespan to prevent them from becoming defective and suddenly harming you. Thanks to constant self-renewal through this process of controlled suicide, cells are prevented from accumulating too many defects over time. However, many cell divisions become less efficient when cells get old, and their repair mechanisms are more error-prone.

One of the main reasons our regenerative potential decreases with age is cell senescence [16]. However, these "retired" cells are not completely inactive. Instead, they continue to produce substances that trigger an unconscious inflammatory response. Accelerated aging and age-related diseases are some of the consequences of these processes. Senescent cells with incurable DNA damage or shortened telomeres make up 20% of cells in older primates [17]. At a certain age, our stem cells also retire and are no longer able to ensure a continuous supply of young cells. The aging process can usually be observed in us humans from our mid-30s onwards, initially through the graying of our hair. This occurs because the stem cells in the hair root, which are responsible for producing the brown pigment melanin, stop working. But as we age, we not only become grayer and wiser—statistically speaking, the risk of developing cancer also increases significantly.

This is also one reason why we are more susceptible to developing cancer as we get older. In old and defective cells, the control mechanisms can be completely disrupted. Copying errors in the genome can disrupt the balance and lead to uncontrolled cell growth, which promotes the formation of tumors and the development of cancer.

References

1. Sender R, Fuchs S, Milo R. Revised estimates for the number of human and bacteria cells in the body. PLoS Biol. 2016;14(8):e1002533. https://doi.org/10.1371/journal.pbio.1002533.

2. Abbott A. Scientists bust myth that our bodies have more bacteria than human cells. Nature. 2016; https://doi.org/10.1038/nature.2016.19136.

3. Der König aller Krankheiten: Krebs—eine Biografie: Mukherjee, Siddhartha, Pleitgen, Fritz, Schaden, Barbara: Amazon.de: Books. https://www.amazon.de/K%C3%B6nig-aller-Krankheiten-Krebs-Biografie/dp/3832196447. Accessed 05 July 2024.

4. Heikenwälder H, Heikenwälder M. Der moderne Krebs - Lifestyle und Umweltfaktoren als Risiko. Berlin/Heidelberg: Springer Berlin Heidelberg; 2023. https://doi.org/10.1007/978-3-662-66576-3.

5. NOVA Online | Cracking the code of life | Genome Facts. https://www.pbs.org/wgbh/nova/genome/facts.html. Accessed 05 July 2024.

6. DNA - Kids | Britannica Kids | Homework Help. https://kids.britannica.com/kids/article/DNA/390730. Accessed 05 July 2024.

7. Dettmer P, Vogel S, Flückiger A. Immun: Alles über das faszinierende System, das uns am Leben hält. Das Immunsystem erklärt vom Macher des beliebten.

8. Eukaryotic Genome Complexity | Learn Science at Scitable. http://www.nature.com/scitable/topicpage/eukaryotic-genome-complexity-437. Accessed 05 July 2024.

9. Kolb JP, Oguin TH III, Oberst A, Martinez J. Programmed cell death and inflammation: winter is coming. Trends Immunol. 2017;38(10):705–18. https://doi.org/10.1016/j.it.2017.06.009.

10. Haas S, Trumpp A, Milsom MD. Causes and consequences of hematopoietic stem cell heterogeneity. Cell Stem Cell. 2018;22(5):627–38. https://doi.org/10.1016/j.stem.2018.04.003.

11. The New Me in 80 Days. The race to replace the cells in my… | by Rich Sobel | ILLUMINATION-Curated | Medium. https://medium.com/illumination-curated/the-new-me-in-80-days-454a3f65409f. Accessed 05 July 2024.

12. Checkpoints bei der Zellteilung | Max-Planck-Institut für Biochemie. https://www.biochem.mpg.de/570404/20021007_nigg_checkpoints. Accessed 05 July 2024.

13. Phases of the cell cycle (article). Khan Academy. https://www.khanacademy.org/science/ap-biology/cell-communication-and-cell-cycle/cell-cycle/a/cell-cycle-phases. Accessed 05 July 2024.

14. Cell Cycle. https://www.genome.gov/genetics-glossary/Cell-Cycle. Accessed 05 July 2024.
15. Medizinwissen, auf das man sich verlassen kann | AMBOss. https://www.amboss. com/de. Accessed 05 July 2024.
16. Khosla S, Farr JN, Tchkonia T, Kirkland JL. The role of cellular senescence in ageing and endocrine disease. Nat Rev Endocrinol. 2020;16(5):263–75. https:// doi.org/10.1038/s41574-020-0335-y.
17. Herbig U, Ferreira M, Condel L, Carey D, Sedivy JM. Cellular senescence in aging primates. Science. 2006;311(5765):1257. https://doi.org/10.1126/ science.1122446.

2

How Cancer Develops

Contents

Abstract After talking about the normal cell, its division and the cell cycle, we move on to tackling the concept of *cancer*. Despite being a daunting word, it is worth understanding what cancer patients deal with. We will also cover examinations and treatments at a later stage. Our goal is to help lessen the "loss of control" and helplessness that cancer patients often feel.

As the human body has a huge number of different types of cells, cancer can start from almost anywhere in the body. Based on its properties, cancer can also spread to other body parts. In the previous chapter, we discussed different biological processes in our cells. In contrast to normal cell behavior, cancer cells grow uncontrollably, and cell division becomes a problem. The normal function of our body during life is to grow and replace old cells with new ones. Healthy cells are not expected to multiply rapidly, except when it comes to building up an army of white blood cells to deal with infection, during repair processes and growth into adulthood. When cells grow old or have been damaged, they die, and new cells take their place. Sometimes, this process does not work correctly, and abnormal or damaged cells grow and multiply when they should not. These damaged cells first form small knots as small as a pinhead (remember, one water droplet contains 500,000 cells). Then, they grow bigger, to the size of a marble, and voilà, we have the tumor.

In this chapter, we will compare normal cells with cancerous cells. We will talk about how mutations evolve and the difference between passenger and driver mutations. What happens when tumor suppressors and oncogenes are out of control? To answer that question, we will dig deeper into the microenvironment, the development of the growth, and metastatic spread. Finally, we will discuss different cancer types, their hallmarks, and their risk factors.

2.1   Cancer Cells vs. Normal Cells

As discussed in the previous chapter, unrepairable errors in the cell cycle can accumulate and lead to cancer cells forming. When a healthy cell becomes cancerous, it loses its specialized function and ceases to become part of the whole system. It refuses to cooperate with other cells and starts to become independent, like a rebel. But things do not end there. Cancer cells continue to strive for more. It is not just a lump in the body but rather a living being inside us that develops. Its primary goal is to migrate, invade organs, and destroy tissues by robbing them of the essential resources they need to live.

Unlike normal cells, cancer cells *refuse to follow signals* for programmed cell death; thus, they cannot *stop growing.* Normal cells can divide several times, but eventually, they exhaust their potential after a few generations of this process. However, cancer cells can *continue to divide endlessly* without any sign of exhaustion. Let's have a small diversion here: More than half a century ago, a young woman named Henrietta Lacks died of an extremely aggressive cervical cancer in the United States shortly after the birth of her fifth child [1]. Shortly before her death, scientists at Johns Hopkins Hospital in Baltimore took tissue samples from her tumor. They were able to successfully multiply them in cell culture. This led to the creation of the first human cell line that is considered "immortal"—the famous HeLa cells. These cells have been used to publish over 75,000 scientific papers and have won four Nobel prizes in medicine [2, 3]. In total, around 50 million tons of HeLa cells have been cultivated to date, which impressively illustrates the remarkable ability of cancer cells to divide [4].

In order to survive, cancer cells require mutations in different types of genes. This change in their genes allows them to overcome the checkpoints that tightly control cell proliferation, which thus enhances their survival and communication with normal cells.

Unlike healthy cells, which are often identical, cancer cells are ***heterogeneous*** or very different from each other. Some cancer cells contain genetic changes that are not found in other cancer cells, thus leading to very different appearances (Fig. 2.1). The shape and coloration of the cancer cells seen under the microscope differ according to their origin. Researchers have taken advantage of these facts by developing therapies targeting the abnormal features of these cancer cells.

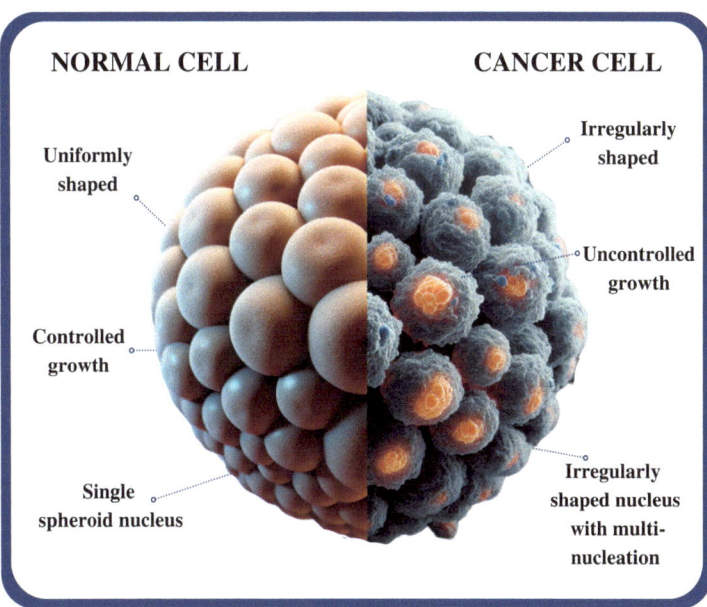

Fig. 2.1 Normal cell vs. cancer cells. Cancer cells have distinct characteristics that set them apart from normal cells

2.2 Tumors

The word ***tumor*** comes from the Latin word "tumere," which means "to swell." It typically refers to a lump or a swollen part of the body. Tumors can originate from any tissue or cell in the body. Since the human body is made up of different types of cells, there are also different kinds of tumors. Each tumor has distinct characteristics, leading to diverse symptoms and disease progression stages.

Tumors are often classified as either cancerous (***malignant***) or not cancerous (***benign***). However, so-called "benign" tumors can become malignant as a result of new mutations. The important difference between benign and malignant tumors is their potential to invade other organs. ***Benign tumors*** displace neighboring tissues without attacking them. They are ***locally limited*** by a shell or capsule or by the natural limit of an organ (e.g., in the original location, carcinoma in the breast is confined in the milk ducts). Benign tumors cannot infiltrate, invade, or spread to other places in the body; they only grow as a cell mass (Fig. 2.2). Benign tumors are thus not referred to as cancer, even

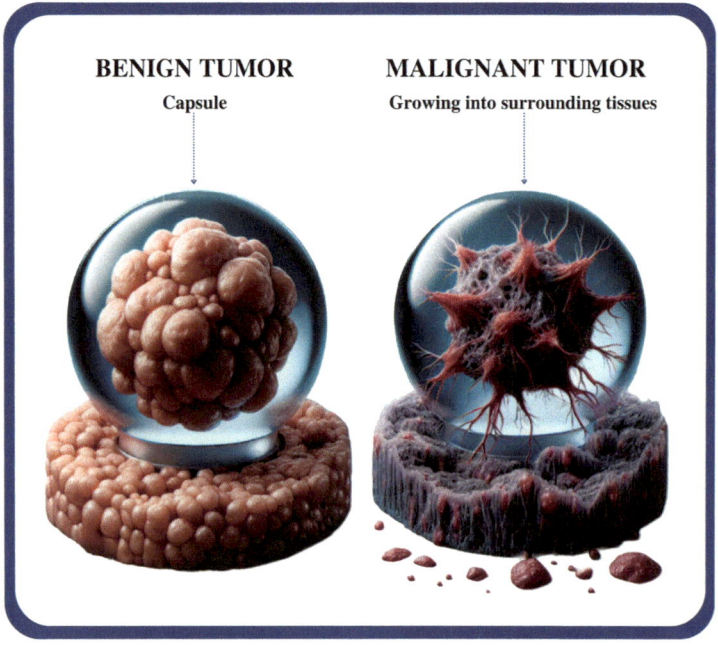

Fig. 2.2 Benign vs. malignant tumors. Benign tumors are encapsulated, which limits their ability to spread. Malignant tumors, however, lack capsules, allowing them to invade nearby tissues and spread through the blood and lymphatic system

if they are sometimes called *carcinoma* (e.g., as we saw above with the example of an "in situ carcinoma" in breast cancer). Because of their volume, they can be problematic for neighboring organs, which can hinder their function. Under the microscope, cells in benign tumors are often seen as identical to the original tissue. They have fewer and less harmful mutations than cells from invasive cancers. They can become massive but grow relatively slowly. While not cancerous, benign tumors can nevertheless be dangerous. They can exert intense pressure on surrounding organs. Furthermore, benign tumors can constrict blood vessels, nerves, and the windpipe by merely expanding, thus leading to death. They can also become more aggressive and turn into invasive malignant cancer. In general, however, benign tumors are much less dangerous than malignant tumors, as we can usually remove them without damaging the surrounding tissues. In fact, most benign cases are considered cured after surgeons remove these tumors.

Malignant tumors, which are commonly referred to as *cancer,* are characterized by rapid growth with unclear or missing tumor boundaries. These boundaries become blurred as the cells infiltrate the surrounding tissue, similar to a plant with small roots. They spread, invade, and attack the neighboring tissue, the blood, and the lymphatic vessels before destroying them. (Fig. 2.2) They can also travel to other places in the body to form new tumors. However, there are some tumors that attack the neighboring tissue but cannot travel to other organs. A typical example of this is a skin tumor called basocellular carcinoma, mainly caused by intense sun exposure. This tumor, although it is malignant, is rarely responsible for metastasis, unlike other skin tumors, such as melanoma, which has a much greater chance of spreading through lymphatics. Melanoma and lung cancer cells contain the most genomic changes, with over 200 mutations per cell, demonstrating the strong carcinogenic influence of UV radiation and cigarette smoke. Smokers' lung cancers include ten times more mutations than those of nonsmokers [5]. As we learn, the behavior of cancer can vary significantly from one organ to another and from one type of cancer to another. Metastases are the most vulnerable malignancies, accounting for 90% of all cancer-related deaths [6]. Despite all the advances in modern medicine, we are still mostly defenseless against them. Removing or radiating the microscopic cancer cells that hide throughout the body is simply impossible; we may not even locate them all. Furthermore, metastases frequently remain in "permanent sleep" for several years, making them resistant to practically all medicines. To tackle them successfully, we need a medicine that is both small and persistent. This is why experts believe the immune system is best suited for this duty. However, we have not yet arrived at our goal. Even though immunotherapies for some malignancies have proven helpful, they are limited. A complete cure is frequently impossible. As a result, the goal of research should not only be to find a cure but also to delay the disease and improve quality of life. Someone who can live well with cancer for many years has already accomplished a lot.

Malignant tumors can lead to complications that affect the entire body. These effects depend on the organ the cancer cells migrate to and the number of cells that invade those organs. For instance, if the cell count is sky-high and the bulk is sufficient, symptoms such as heavy night sweats, body weakness, and unintentional weight loss may arise. However, some cancers develop slowly at the beginning, and because they progress relatively slowly, early-stage cancer enables opportunities for intervention and prevention.

Just as there are two types of tumors, there are two basic types of cancer. A **solid cancer** is formed in solid tissues, such as the muscles, bones, and organs. On the other hand, **cancers of the blood,** such as leukemia, myeloma, and lymphomas, affect the blood, bone marrow, lymph, and lymphatic circulation. They originate from the bone marrow and extend to where the major "highways" of the blood, the vascular and lymphatic systems, are situated.

2.3   Mutations and Cancer

As we know now, unchecked cell growth and proliferation cause cancer. Because of uncontrolled cell division, DNA is copied multiple times, resulting in occasional copying errors. Moreover, **repair mechanisms** either **work less** or **not at all** because of their degeneration. Due to these problems, notably in DNA repair, genes may be expressed as mutations. A mutation is a **genetic variation** that may increase the risk of developing a disease or cause the disease itself. Once a mutation happens in the DNA, it can permanently change its function. The **DNA changes** enable the malignant tumor to overcome the limitations of healthy cells. Disruption of the cells' normal functioning plays a crucial role in cancer development. Some genes are so important in cell division that if they are mutated, patients are at high risk of developing certain cancers.

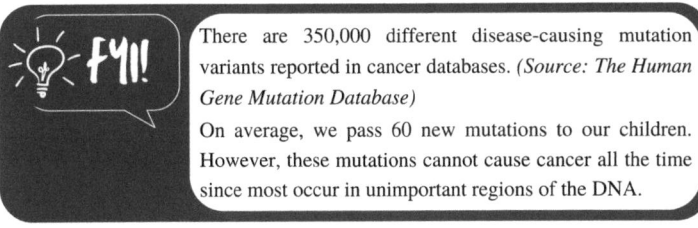

There are 350,000 different disease-causing mutation variants reported in cancer databases. *(Source: The Human Gene Mutation Database)*
On average, we pass 60 new mutations to our children. However, these mutations cannot cause cancer all the time since most occur in unimportant regions of the DNA.

Therefore, the altered DNA is the source of carcinogenesis. For instance, if a person has a so-called cancer gene, it indicates that they have a gene mutation that makes them more likely to get cancer.

We can compare creating mutations to creating books by using a ***photocopy machine*** (Fig. 2.3). Most of the time, copies are made correctly. However, when done repeatedly, we cannot avoid mistakes. For instance, a page can have duplicates in one book, while in others, some pages go missing. These mistakes impact the information in the book. Over time, these mistakes accumulate or occur in abundance, which could significantly affect the overall content of the copied book. Sometimes, there are serious consequences, and the book becomes incomprehensible, as described in "The Story of the Siege of Lisbon" by José Saramago, winner of the Nobel Prize for Literature, which recounts how a proofreader at a publishing house decides to replace a "yes" with a "no," thus changing the entire history of Portugal.

Mutations have many causes. They can occur spontaneously as a result of faulty DNA repair or an error during cell division. These are known as acquired mutations. They are only found in cancer cells and not in other cells of the body. On the other hand, there is another kind of mutation that is passed on from parents to offspring: hereditary mutations. These are found in all the body's cells (and not just in cancer cells). The latter part of this chapter will discuss how these factors lead to cancer-causing mutations.

Returning to our cookbook analogy, there may be moments when we unintentionally alter the ingredients or their measurements while cooking despite following the recipe correctly (Fig. 2.4). These slight changes might make the dish taste better, ruin it entirely, or not affect the dish's flavor at all. Similarly, mutations affect cells differently. The cell could remain unchanged, become linked to another disease, or become a cancer cell.

Although many mutations decrease the chance of survival, others can also improve adaptation and give the cell a decisive advantage in survival by making the cell stronger and more resistant. This is because nature accepts the death of a single individual by constantly experimenting with mutations to enable the survival of the whole species through adaptation. Therefore, mutations are necessary. Without them, there would be no development and evolution and thus no *Homo sapiens* (human beings).

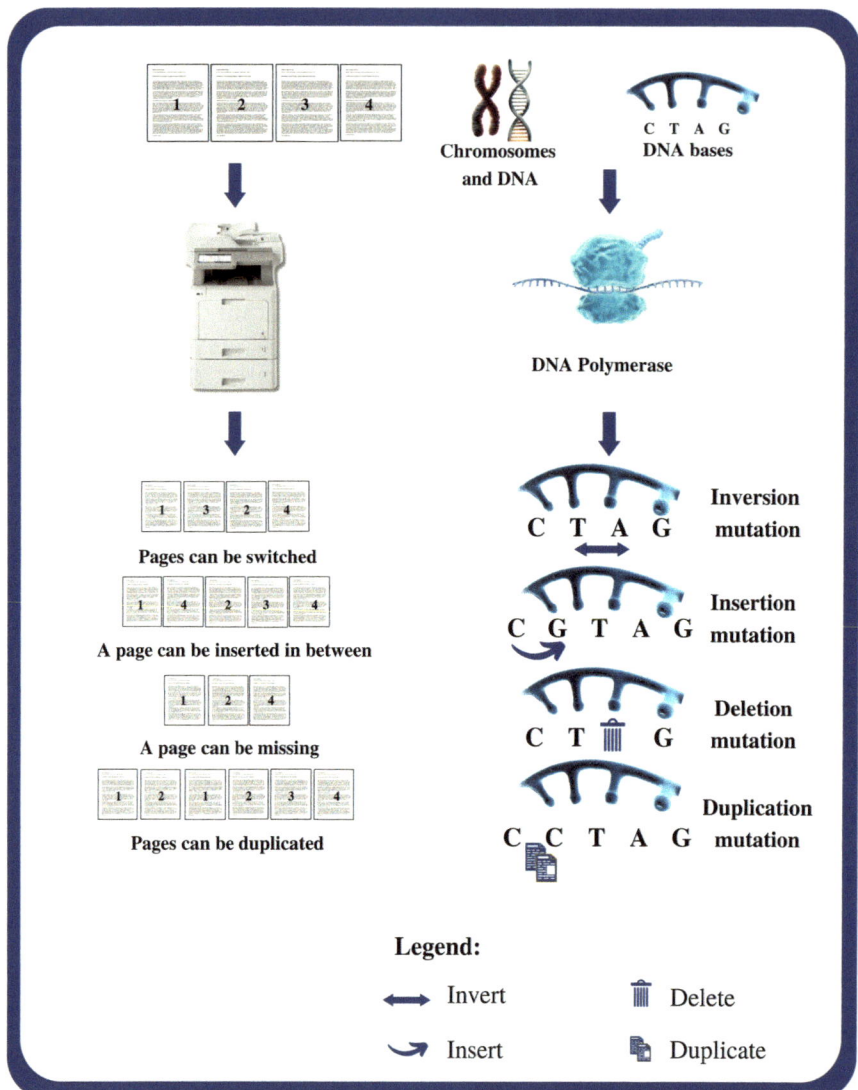

Fig. 2.3 Analogy types of mutations. When using a photocopying machine to create multiple copies of a book, mistakes will inevitably occur. Pages can be switched, inserted between other pages, or be missing or duplicated. Similarly, in cell replication, different types of mutations can occur, such as inversion mutations (where DNA bases are switched), insertion mutations (where a DNA base is added), deletion mutations (where a DNA base is randomly removed), and duplication mutations (where DNA bases are copied)

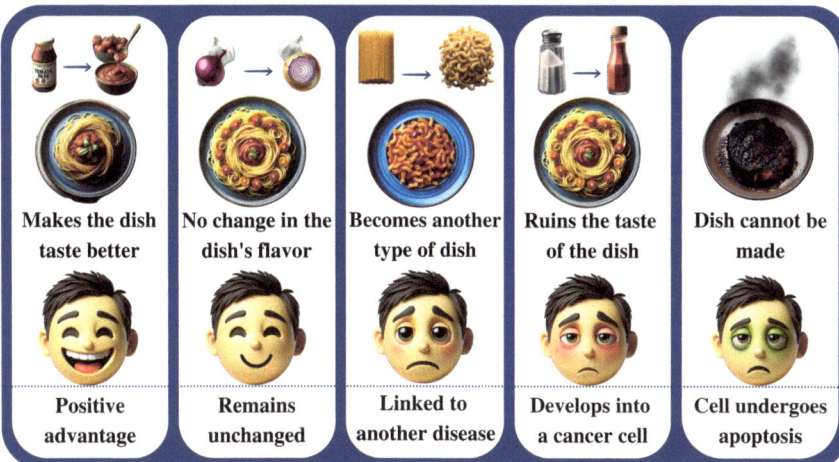

Fig. 2.4 Effects of mutations. Changing the ingredients in a recipe can significantly impact the final dish. Similarly, mutations can bring about a range of effects on cells, including positive changes, no change at all, or negative consequences, such as the development of cancerous cells

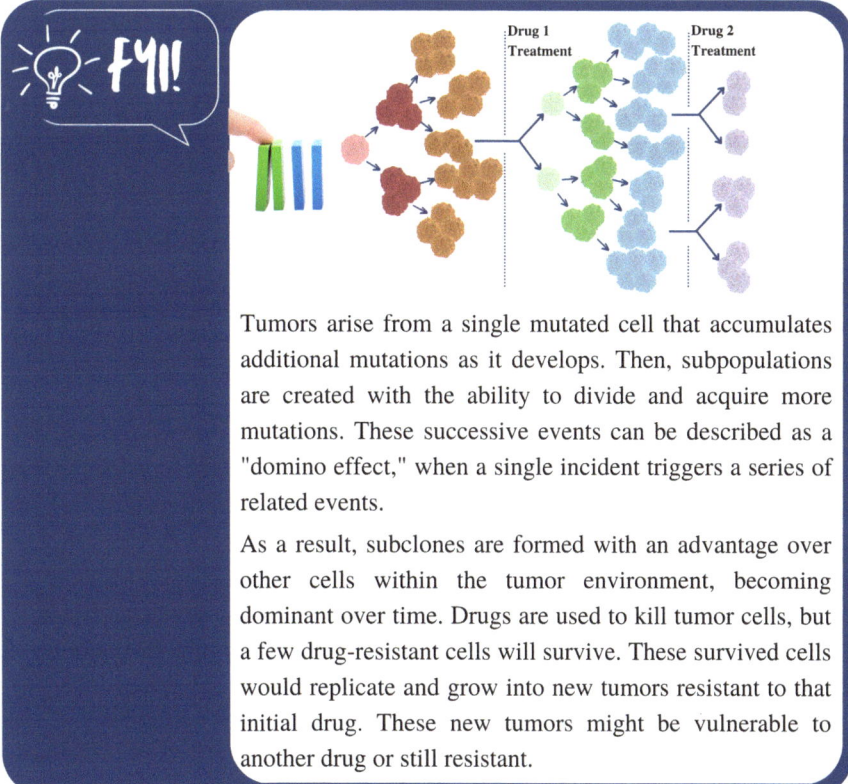

Tumors arise from a single mutated cell that accumulates additional mutations as it develops. Then, subpopulations are created with the ability to divide and acquire more mutations. These successive events can be described as a "domino effect," when a single incident triggers a series of related events.

As a result, subclones are formed with an advantage over other cells within the tumor environment, becoming dominant over time. Drugs are used to kill tumor cells, but a few drug-resistant cells will survive. These survived cells would replicate and grow into new tumors resistant to that initial drug. These new tumors might be vulnerable to another drug or still resistant.

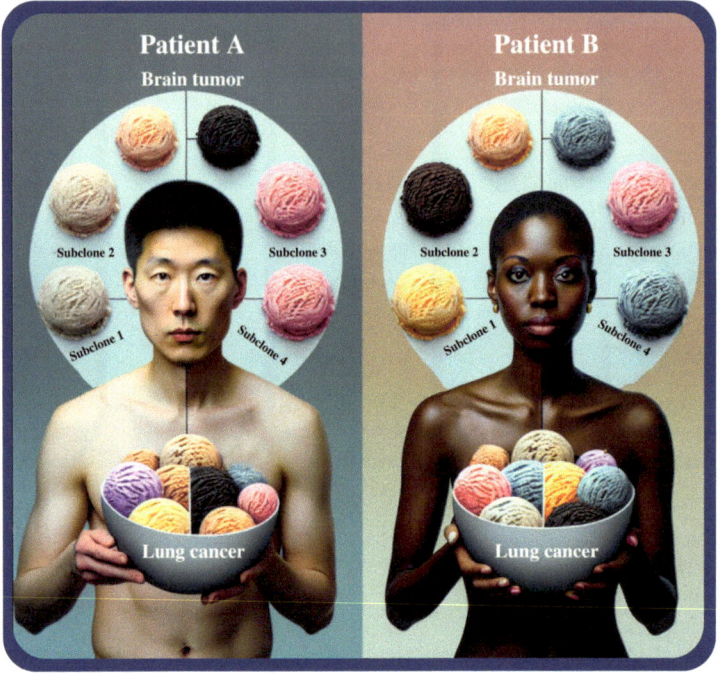

Fig. 2.5 Cancer heterogeneity. Cancer cells within a tumor can be composed of diverse subpopulations with distinct genetic mutations. Furthermore, tumors within an individual can also exhibit significant variability. This diversity in cancer can be likened to the variety of flavors and presentations found in ice cream

Mutations are the main reason why cancer is a very *heterogeneous* disease. In other words, a tumor may be composed of a combination of several types of cancer cells and not just a single species of cancer cell. Each generation of cancer cells can have new mutations. Therefore, no two individuals have the same cancer, and even tumors within a single individual are likely to be very different.

This heterogeneity is the main reason why cancer is very difficult to treat. For example, (Fig. 2.5) shows that Patient A and Patient B have been diagnosed with colon cancer. However, the tumors in their colon contain different subclones (subpopulations with genetic mutations).

The colon tumors contain different "clones." A "clone" is a population of cells with similar genetic mutations. In the same cancer, there are usually several clones, i.e., several different groups of cells. Within each group, the cells resemble each other. In this example, the patients' cancers have metastasized to the brain. The metastases may differ from the original tumors (in the colon). A clone of the original tumor has moved into the brain and acquired

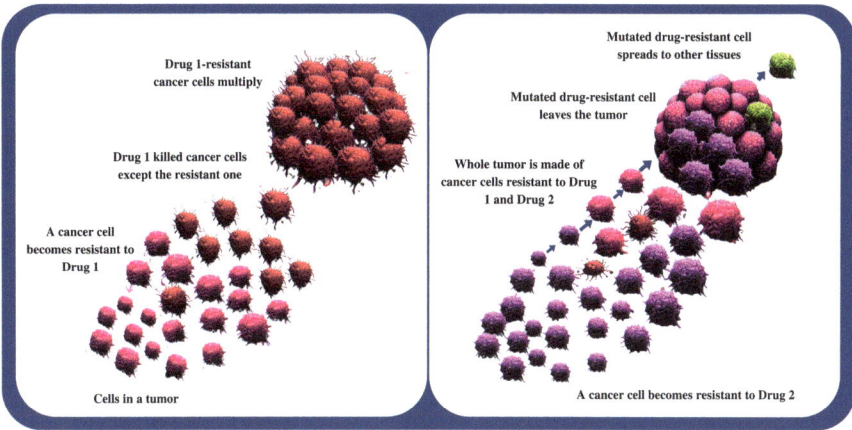

Fig. 2.6 Drug-resistant clones. A drug-resistant cancer cell can exist within a tumor that survives drug treatments. This drug-resistant cancer cell can duplicate and multiply, creating a tumor that can resist the drug treatment. In addition, it can also induce other drug-resistant mutated clones

new mutations "on the spot," which explains this difference. On the other hand, they also have metastasis in their brains. These tumors are also very different from the tumors in their colons. This heterogeneity is like ice cream. It is a single type of food with various flavors, made up of different ingredients, and it can be served in different combinations.

When a drug or the body's immune cells attack cancer, most of the time, they destroy the most vulnerable cancer cells, thus giving resistant cancer cells a chance to develop (Fig. 2.6). The mutated clones that arise are cancer cells that have adapted to survive external pressure and are now resistant to another attack. The constant selection pressure, new mutations, and the proliferation of these cells lead to these cells adapting, growing, and thriving. In some cases, mutations also promote new mutations, creating even more mutant clones [7].

The speed at which cancer develops does not depend on the patient's age. It depends on the patient's immune defenses but, above all, on the characteristics of the cancer cells, which is the type of mutations they have. For example, cells that have developed a mutation in the apoptosis gene (the gene responsible for the voluntary death of cells, known as BCL2) can no longer die, even if they are very old or damaged. In this case, the tumor progresses slowly. If, on top of this, a mutation affecting a proliferation gene is added (e.g., the MYC gene), the situation becomes catastrophic, as the tumor cells begin to divide very rapidly while being unable to die. In this case, the tumor develops extremely rapidly, and we'll come back to this subject in later chapters.

2.4   Driver and Passenger Mutations

Some mutations occur in the DNA segments that are vital in repairing genetic material, so they are called ***driver mutations*** [8]. Because driver mutations escape repair and thus remain in the genome of the affected cell, this can be the first step to developing cancer. They are passed on with each cell division and affect crucial oncogenes and tumor suppressors, which will be explained in the next section.

Due to cancer heterogeneity, a single-driver mutation is not enough to alter tumors. Instead, mutations build up over time, and they, in turn, alter tumors. This process is like driving a bus. The driver can control where the bus goes and how fast it travels. If anything happens to the driver, the bus can go out of control or stop. During the trip, external factors also influence the structure of the bus. For example, the bus might suddenly burst a tire, making it go slower than usual. The weather can also significantly impact the bus. For example, hail can destroy the headlamp and windshield and thus reduce visibility. In that case, the driver may not see a red light and thus drive into another car. When these external factors accumulate, the bus can break down and never reach its final destination (Fig. 2.7).

On the other hand, some mutations remain insignificant, even without repair. They are called ***passenger mutations***. These mutations are mainly caused by copying errors that occur during cell division and in unimportant sections of the DNA. They are spread indiscriminately, remain in the genome, and are passively passed on with each cell division. Furthermore, they are identifiable but inconsequential. This is like a bus ride. Inside the bus, the actions of the passengers accompanying the driver may not affect the motion of the bus very much. Other external factors, such as the weather, for example, may have little to no effect on the structure of the bus (Fig. 2.7).

Normal cells randomly develop driver and passenger mutations. This marks the events of early tumorigenesis. While a cell does not "forget" any mutation, driver mutations are harmful mutations that can accumulate and lead to cancer. Each cancer cell has a specific number and combination of driver and passenger mutations. Driver mutations in the genome are limited in number. For example, a colon cancer patient could have 130 mutations. However, only about ten may have contributed to the tumor's growth and impacted its survival. Those ten are the driver mutations; the rest are passenger mutations spread across the genome waiting to be passed down. Unfortunately, although these two types of mutations function differently, it is difficult to tell them apart.

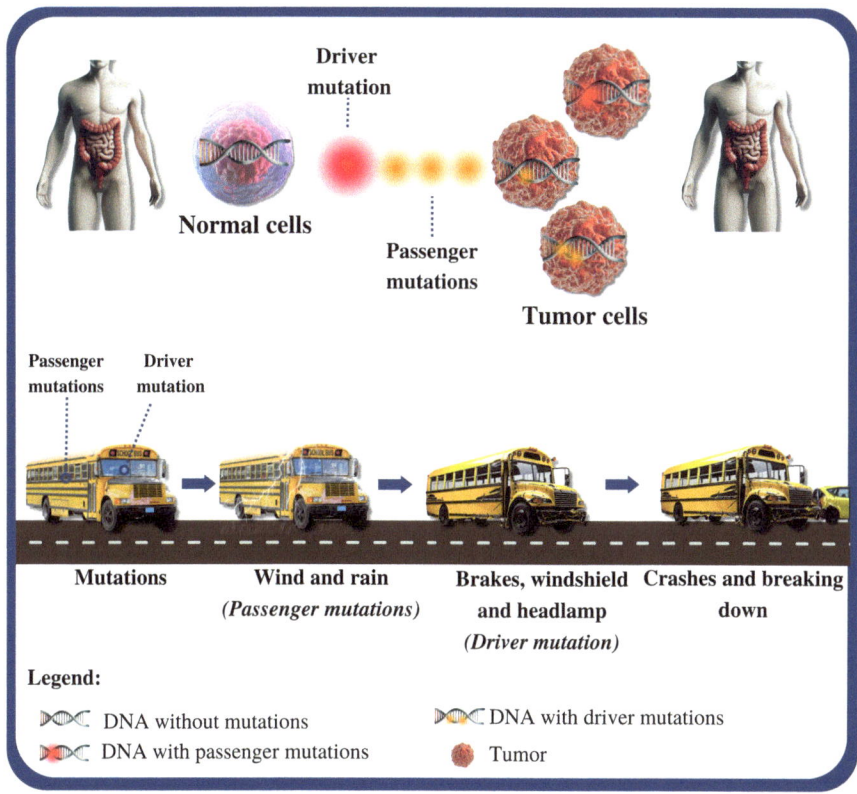

Fig. 2.7 Driver and passenger mutations. Driver mutations cause normal cells to become cancer, while passenger mutations do not. Similarly, problems with the driver or the vehicle, such as broken brakes, windshields, and headlamps, can have serious consequences and lead to an accident. In contrast, problems with the passengers and the weather, such as wind and rain, may have little to no effect on the bus

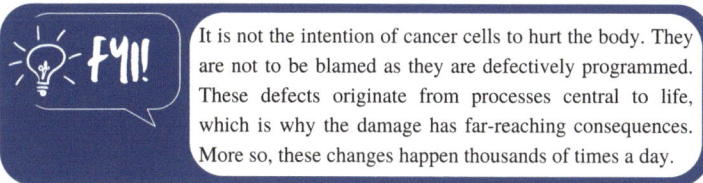

It is not the intention of cancer cells to hurt the body. They are not to be blamed as they are defectively programmed. These defects originate from processes central to life, which is why the damage has far-reaching consequences. More so, these changes happen thousands of times a day.

2.5   Tumor Suppressor and Oncogenes

There are two types of cancer genes: tumor suppressor genes and proto-oncogenes. Let us start first with the ***tumor suppressor genes***. These genes are the gatekeepers of normal cell function that protect us against tumors by controlling normal cell division and preventing tumor development. They are responsible for repairing damaged genetic information and generating safety and control mechanisms that constantly search for deviations and copy errors in the DNA to repair them immediately.

However, inactivating mutations reduce tumor suppressor gene functions. Tumor suppressor genes lose their ability to repair themselves. To understand this, consider the most famous tumor suppressor, **TP53,** which actively suppresses the uncontrolled division of genetically damaged cells. If mutations inactivate tumor suppressor genes, the "stop" signal marking the end of cell division is no longer registered. As a result, cell division will remain uncontrolled and unregulated [9]. Humans have two copies of tumor suppressor genes in every cell. For the inactivation to take effect, both copies of the tumor suppressor gene must be mutated. In the case of elephants, they have multiple copies of *TP53,* which is one of the main reasons they rarely get cancer. We can imagine *TP53* as the brakes of a car. If we go too fast, we can hit the brakes and stop. If the brakes are broken, however, the car cannot stop and will crash (Fig. 2.8). *TP53* is a brake that works to suppress tumors.

An example of an **oncogene** is a gene called *Src*. It accelerates cell division upon receiving an appropriate growth signal. When mutated, *Src* is driven to a state of constant hyperactivity, which leads to uncontrolled cell division. Oncogenes are like gas pedals in a car; if they get stuck, the car will accelerate beyond control. Similarly, a cell cannot stop dividing if the *Src* gene is mutated (Fig. 2.8). Out of all our genes, only **70 human oncogenes** have been identified and linked to human cancers. These also include *RAS*, which we will discuss later with a story about a 45-year-old fire protection technician.

On the other hand, a crucial mutation may also occur in an oncogene. An oncogene is a gene responsible for the growth, division, multiplication and specialization of cells. An oncogene is active during the embryonic stage of human development. This gene enables a single cell to become trillions in a short timeframe. When there are enough cells for a complete human being to be made, the oncogene is switched off and becomes a proto-oncogene. Even when only one of the two gene copies undergoes alteration, mutations in oncogenes enhance cell proliferation. Oncogenes are rarely inherited due to their excessive activity, which can disrupt finely coordinated embryonic development and lead to early miscarriages. Instead, mutations in oncogenes that promote cancer cell growth and survival are more likely to occur later in life [10].

However, decades later, these oncogenes may be switched on again due to mutations. This switching-on process promotes cancer and disrupts

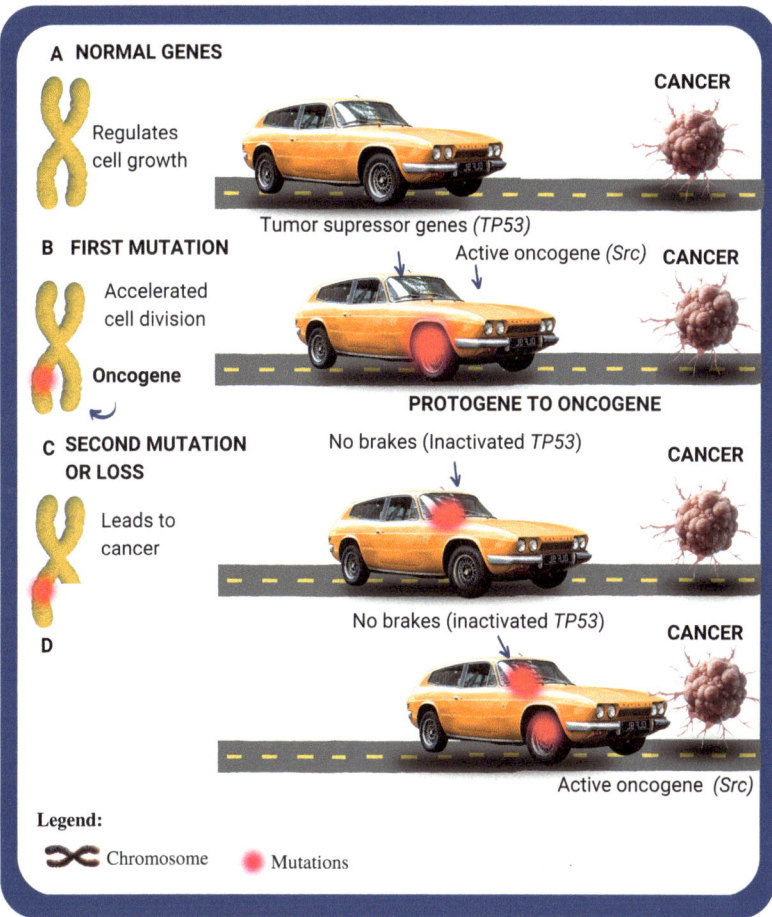

A NORMAL GENES

Regulates cell growth

Tumor supressor genes *(TP53)*

CANCER

B FIRST MUTATION

Accelerated cell division

Oncogene

Active oncogene *(Src)*   CANCER

PROTOGENE TO ONCOGENE

C SECOND MUTATION OR LOSS

Leads to cancer

No brakes (Inactivated *TP53*)

CANCER

D

No brakes (inactivated *TP53*)

CANCER

Active oncogene *(Src)*

Legend:

Chromosome   Mutations

Fig. 2.8 Cancer cells develop in steps. (**a**) The cell grows and normally proliferates when both proto-oncogenes and tumor suppressor genes are regulated. (**b**) The cell cannot stop dividing when a proto-oncogene is transformed into an oncogene. (**c**) Unlimited cell division occurs if mutations inactivate tumor suppressor genes yet activate oncogenes. (**d**) As a result, cancer starts to develop

regulatory controls. This means oncogenes are expressed with a new function at the wrong time or level. As a result, the cell can again multiply very rapidly.

Tumor suppressor genes and oncogenes interact in various ways. Their interactions can be understood by thinking of a car driving on a road. To drive a car, we need an accelerator (gas pedal) and brakes, among other things. This is similar to normal and healthy genes, which have many genes to suppress tumor genes but also a few that help tumors grow (oncogenes). Mutations of the tumor suppressor and proto-oncogenes can occasionally occur in the same cell over a lifetime. Those two steps can happen independently, and the time interval between these two events can sometimes be as long as a decade. Going back to our car analogy, this would mean that if the accelerator is hyperactive now and the brakes do not work later on, then an accident is unavoidable. Translating this into biology means that cancer cells have many oncogenes but few tumor suppressor genes. As a result, the cursed cell cannot stop dividing, and cancer develops. Cancer cells have many oncogenes but few tumor suppressor genes.

In earlier discussions, we covered methylation, a process where genes are locked, hindering their proper functioning. When a tumor suppressor gene undergoes methylation, it becomes locked and can contribute to tumor development. For instance, methylation of the *BRCA1* gene, a tumor suppressor gene, can increase the risk of breast or ovarian cancer in certain individuals, even without a mutation in the gene.

2.6   Signaling Pathways

Proteins are major components of a cell. They play a vital role in cellular function, serving as key building blocks. Their unique shapes allow for precise interactions, either with other proteins or within themselves. These interactions can convey important messages within the cell, such as signals for proliferation. The sequences of interlocking and shape-shifting events that pass from one protein to another to transmit these messages are known as ***signaling cascades*** or ***pathways***. They are tightly regulated to enable cells to perform certain functions.

How does a signaling pathway work? A cell receives information from its environment through the receptors on its surface. The receptor is activated when the signal is accepted and is brought inside the cell. The proteins transmit these signals by changing their shapes. This change activates the next protein to change shape, which will activate another protein, leading to a cascade until the last protein in the pathway is activated and the intended cell function is carried out [11].

In simpler words, a cell gets a signal from the outside. It forwards it to its final receiver, the nucleus. This causes a biological response by activating its genes and making the corresponding protein function (biochemical events). In addition, proteins can also switch certain genes "on" and "off." This switching action dictates what the cell can or cannot do and how it responds to certain stimuli.

Proteins are the basic building blocks of all living things, such as plants, animals and humans. They are highly efficient and versatile, and can be used for various things, from sending signals and building simple to complex proteins. They are composed of chains of amino acids, which causes their shape to constantly change. There are twenty different types of amino acids that can be combined in different ways. However, the majority of these possible proteins are not useful.

Multiple signaling pathways can be interconnected. When that happens, one component can receive signals from several inputs, which activates several other signaling pathways, each of which activates other pathways and so on and so forth. The signal will eventually become strong enough to enable a cell to change its "state," such as from one that is not dividing to one that is. However, one might wonder, with all these complex networks interconnected, how do they properly coordinate their actions?

These pathways are like interconnected flights with a synchronized schedule. If one flight is delayed, the next flight will also be delayed, and so on. To maintain order, plane pilots must stay on schedule and follow instructions from air traffic controllers (Fig. 2.9). The same situation occurs in cell signaling: everything is coordinated. Different protein kinases with different entryways help regulate different functions.

What is more, the flow and the duration of the signals are also tightly regulated throughout the network [12]. However, in the human body, this chain of signals can be interrupted if the proteins responsible for their transmission are not functioning correctly.

For instance, if their shape is abnormal. An abnormal shape can prevent them from functioning properly. This can happen if there is a mutation in the gene responsible for their manufacture: the "gene recipe" is poorly written, leading to a protein that is poorly made, has an abnormal shape, does not fit together properly, and therefore functions inadequately. Certain mutations in the EGFR (epidermal growth factor receptor) gene in lung cancer result in the production of an abnormally shaped protein (the protein is epidermal growth factor receptor). Whereas this protein should only send proliferation information to the nucleus when stimulated, it sends this information continuously, contributing to the development of cancer.

Mutations in specific genes of transmission proteins, known as protein kinases, can be the reason behind this disruption. In some cases, these mutations may not have any effects if the protein is not significantly altered.

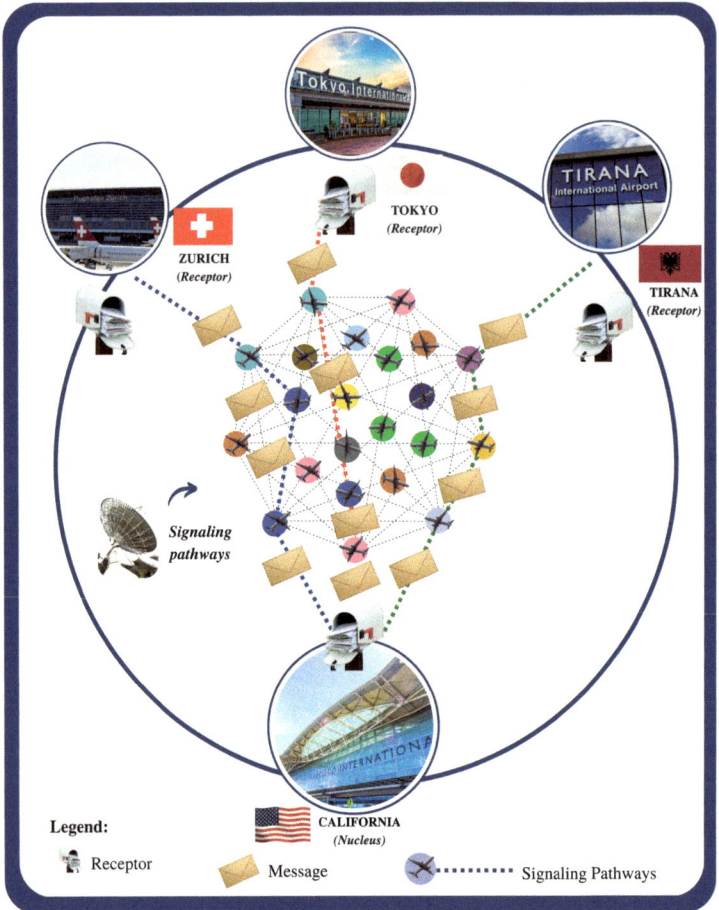

Fig. 2.9 Signaling pathways. In this scenario, Zurich, Tokyo, and Tirana airports serve as pivotal points for originating letter messages. These messages seamlessly traverse interconnected flight networks to reach their intended recipient in California

However, if the protein is modified, it can cause the protein to stop functioning correctly, which can lead to the transmission of messages being prevented or sent continuously. For instance, if messages encouraging proliferation are continuously sent, it may contribute to the development of cancer. Similarly, if a signal leading to cell death is blocked, the cell will not die, even if it is highly abnormal.

Around *11–15 signaling pathways* are typically disrupted in cancer cells. Despite genetic tumor heterogeneity and diversity, the same signaling pathways are disrupted no matter the type of tumor. One example is the *Ras-Mek-Erk* pathway, which strictly regulates cell division. *Ras* activates the *Mek* proteins, which activates *Erk*, thus accelerating cell division. In cancer cells, activations of these proteins are persistent, leading cells to divide uncontrollably. These hyperactive signals force cancer cells to divide, grow, and become dependent on permanently activated signaling pathways. Another signaling pathway is activated to block cell death signaling, making cancer cells escape death. A different pathway enhances cancer cell survival in hostile new environments. These activated signaling pathways aim to keep cancer alive. Researchers working on cancer treatment drugs target these signaling pathways to prevent cancer cells from growing. The idea is to disrupt the unhealthy signaling pathway and stop cancer from developing. For example, drugs called "tyrosine kinase inhibitors" are used in certain lung cancers whose protein kinases show abnormalities related to certain very specific mutations.

2.7 How Cancer Progresses: A Four-Step Hypothesis

Cancer does not affect a person suddenly when they reach retirement age. Over the years, *mutations* occur gradually and quietly. The number of mutations we accumulate is determined by the number of carcinogens we are exposed to in the environment during our lives or the number we voluntarily expose ourselves to. The number also depends on the genetically determined function of our cellular detoxification mechanisms and chance. The more frequently and the longer a person is exposed to carcinogens, the higher their risk of developing cancer.

Usually, cancer develops in steps that provide the body with plenty of opportunities for prevention and repair (Fig. 2.10). According to one hypothesis, there are four steps in the process. The first step in developing cancer is called *initiation*. During this step, crucial mutations that affect the cell cycle (division, growth, death, or repair and control) appear in the DNA. After these initial mutations occur, further steps in cancer development can take place. While so-called initiated cells are not yet activated, they can proliferate faster than normal cells and form tumors due to constant exposure to toxic agents.

The second step in developing cancer is called *promotion*. This is when the initiated cells become transformed cells. Further mutations in oncogenes and tumor suppressor genes accumulate at the target site in this step. The mutated cell then divides uncontrollably and has a growth advantage over the other cells. The daughter cells then have the same DNA alteration. During the

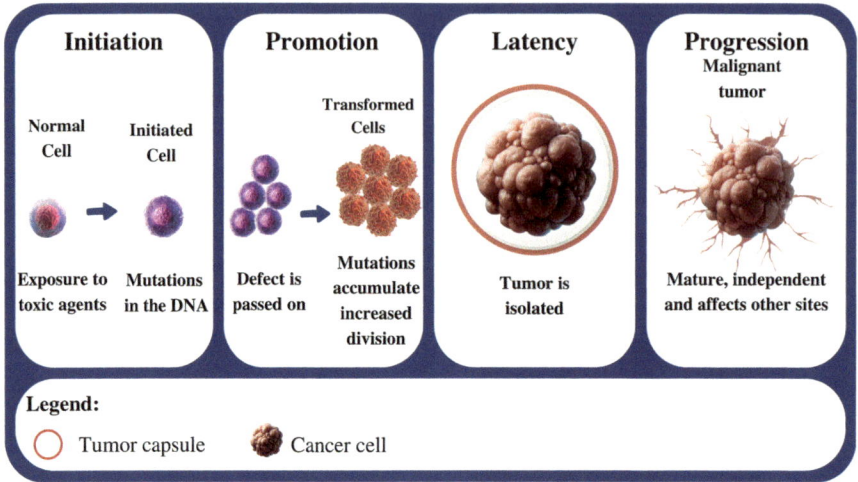

Fig. 2.10 The steps in developing cancer. Cancer development is a gradual process that involves four steps. This process gives us plenty of opportunities for prevention and repair

promotion, microscopic tumors around *0.5 mm³* in size start to appear, with a cell mass of approximately *5000 cells.* In step two, tumors are harmless and do not grow because they lack blood vessels for nutrition [13].

The third step is called *latency*. This is when the body uses a defense mechanism. It tries to cut off the oxygen supply of these microtumors by isolating them and covering them with healthy cells. This is also when further mutations of important genes *accumulate* in the cells. A single mutation can never lead to tumor development on its own; instead, more than one mutation, at least in the other copy of the DNA, is necessary for this to happen. In step three, prevention through proper nutrition, diet, and physical activity can be crucial. It is much easier to eliminate these microtumors or prevent them from developing than to deal with a mature tumor. Specific molecules in food (such as those that will be discussed later) can help keep these microtumors dormant. However, mutations can also occur *spontaneously* for no reason during replication. Such a cell starts to divide uncontrollably, which is why even people with a healthy lifestyle develop cancer.

During step three, latency, pathologists can confirm the presence of microtumors in our bodies. For example, coils of noninvasive precancerous cells can be observed in the uterus tissue before becoming invasive cancer. Similarly, precancerous cells in the lungs of smokers can be found long before the onset of lung cancer. Moreover, premalignant polyps (growth of tissues) are present in the intestines before becoming a carcinoma (these points will be further discussed in the chapter on early detection tests). Interestingly, some patients who were biopsied after dying from non-cancer causes were found to have microtumors in their tissues that were not previously detected. For instance, Asian men with a lower rate of prostate cancer have been found to have the same amount of precancerous microtumors as Western men, even though higher rates of cancer are reported in Western countries. These results indicate that diet and lifestyle can be major factors in whether microtumors in step three turn into cancer.

After some decades, step four may begin as the tumor enters the *progression* phase (Fig. 2.10). This is the final step and involves the malignant transformation of the cell. This transformation marks the moment when the cancer cell has now acquired all the properties it needs due to the many unrepaired DNA mutations that occurred before. During this point, the microtumors develop blood vessels. Since they have an ample supply of oxygen and nutrients, the tumors can continue to grow exponentially [14], up to *16,000 times* bigger than their original size.

In this context of exponential cell growth, we also have uncontrolled cell death in the center of the tumor. One reason is that the high mutation rates in the cell itself suppress the controlled cell death process and initiate necrosis. This is the big dark black bulk we can see in the center of the malignant tumor. Some theories suppose that the cells inside cancer do not get enough food and oxygen that they need to survive because microvessels surround the tumor from the outside. However, the main problem with uncontrolled cell death here is that it releases harmful products into the extracellular space, which initiates inflammation in the surrounding tissue.

Transformed cells in step four become independent, invade local tissue, and spread throughout the body via the blood vessels. This process is called *metastasis*. Metastasis is a Greek word that means "removing, removal, migration, changing, change or revolution." This word refers to cancer spreading from its primary site to more distant organs. The metastatic process can be likened to a tree planted near a river. When a seed drops into the river, it will be carried far away from where it entered the river. Upon finding suitable soil, this seed can germinate and grow into the same plant it came from (Fig. 2.11).

Similarly, a cancer cell can break away from the original tumor, migrate into the nearby blood or lymphatic vessel, settle, and survive in the new tissue and then multiply. The new tumor is of the same type and has the same characteristics as the parent tumor. For this reason, the cells we find in liver metastasis of stomach cancer are not liver cancer cells but stomach cancer cells.

During metastasis, cancer spreads widely, develops daughter tumors, and becomes detectable. Additionally, the risk of further mutation increases in the

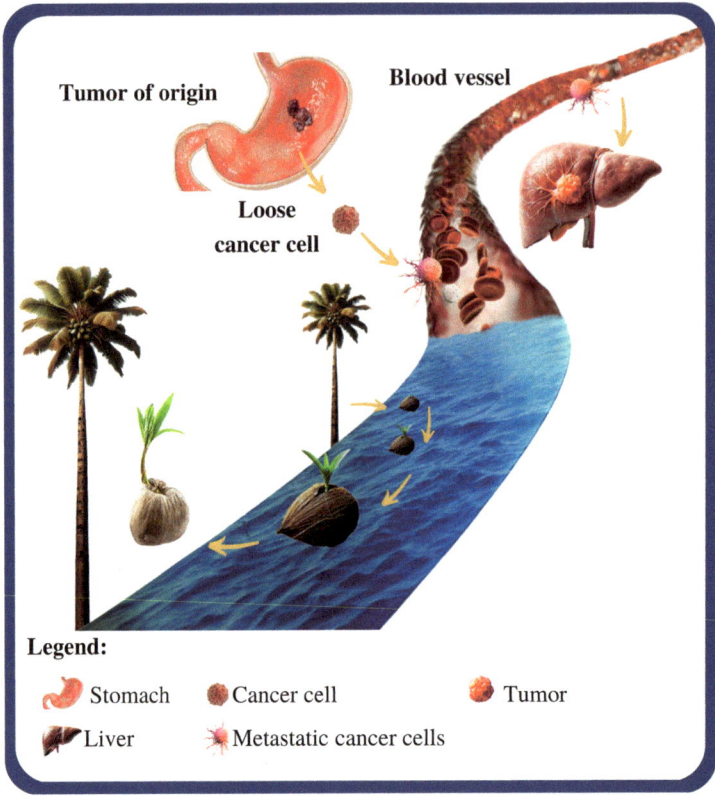

Fig. 2.11 Metastasis. Cancer cells can detach themselves from their original tumor and spread to other organs via blood vessels. Similarly, a coconut that falls into a river may grow into a palm tree once it arrives in a new place [15]

course of the disease in step four as degeneration becomes more probable. In fact, metastatic cancers are much more lethal than non-metastatic cancers.

To conceptualize tumor progression in a real-life situation, let us follow the story of Tom, who had lung cancer. This scenario is excerpted from Siddhartha Mukherjee's book *The Emperor of All Maladies: A Biography of Cancer* (2010). At first, Tom was a 45-year-old man who accidentally inhaled asbestos dust while working. Asbestos is an agent known to cause cancer; it is a ***carcinogen*** (Fig. 2.12). The asbestos dust entered and settled in his left lung. Because it was a foreign substance, the body fought back with inflammation. The cells around the dust particles started to divide uncontrollably, forming a small lump. Randomly, a mutation occurred in the *Ras* gene of one of the cells in the lump, resulting in an activated version of *Ras*. This mutation caused the cell to divide more rapidly, creating a clump of cells within the original lump.

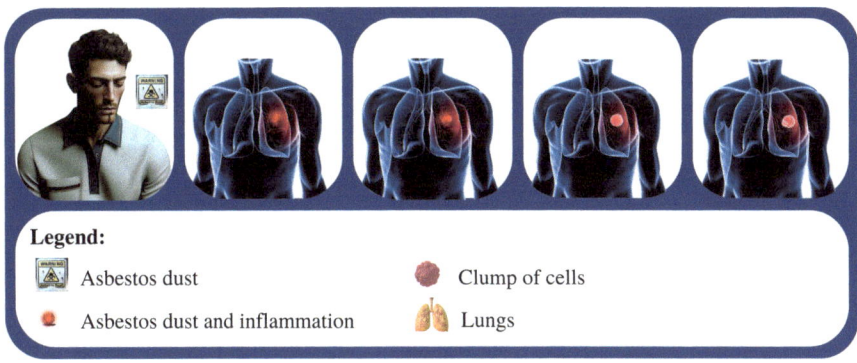

Fig. 2.12 The case of Tom. A foreign particle enters the lung and causes inflammation. This inflammation would lead to the formation of cancerous tumors

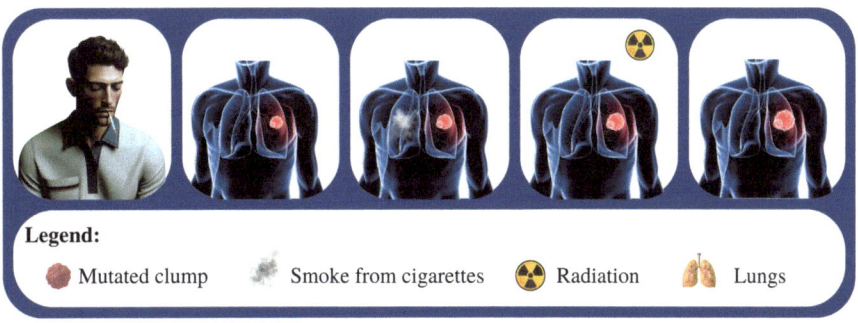

Fig. 2.13 External factors such as cigarette smoking and radiation cause the cancerous tumor to develop more mutations, speeding up its growth

This clump of cells with the *Ras* mutation continued to proliferate unnoticed for a decade [9].

One day, while Tom was smoking, a chemical from the cigarette entered his left lung and encountered a clump of *Ras*-mutated cells (Fig. 2.13). This caused another mutation that activated a second oncogene. Still, the effect of these events remained undetected. Then, some time passed, and a radioactive beam hit the clump one day, which led to a mutation that inactivated a tumor suppressor gene. But it had little effect then since another unmutated gene copy existed. However, after a year, another mutation inactivated the second tumor suppressor gene. This final step launched cancer. At that point, the cell had acquired a total of four mutations and was on the path to unlimited cell division. The cancerous lump grew faster than the nearby cells, so it could be called a tumor [9].

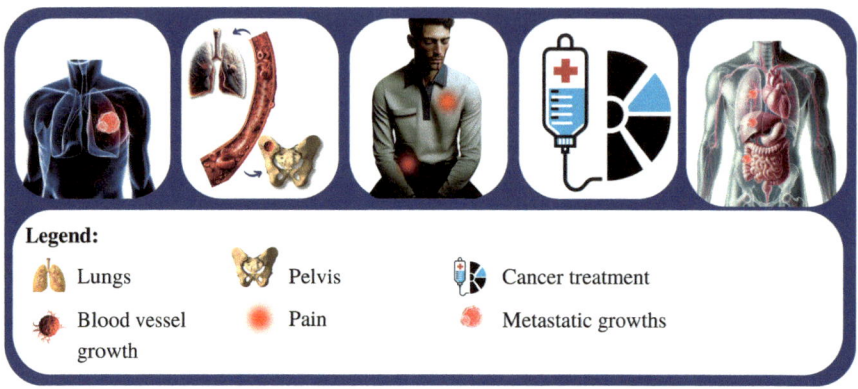

Fig. 2.14 A cell from the cancerous tumor can get loose and grow in other tissues, spreading the disease throughout the body

As the tumor grew, it acquired more mutations and activated multiple signaling pathways. This made the rogue cells more eager to grow and more capable of survival. One of the mutations enabled this tumor to stimulate the growth of blood vessels toward itself (Fig. 2.14). As this tumor became well-nourished, it was able to make copies of itself safely. During replication, it acquired another mutation that enabled it to grow, even in oxygen-poor areas of the body. One cell in the tumor also acquired mutations that improved its motility. This meant that the cancerous cell could travel from the lungs to the bloodstream. Once let loose, that cancerous cell then settled in the bone, where it started another clump of cells that had the harmful mutations, too [9].

One day, Tom experienced several symptoms, such as shortness of breath, tingling pain at the edge of his lungs, and a moving sensation under his rib cage. Within a year, these symptoms started to intensify. He went to see a doctor. He was diagnosed with lung cancer, with a mass wrapped around his left bronchi. After a few weeks, Tom then felt pain in his ribs and hips. A CT scan revealed metastatic growth in these places. Since the tumor in the lung was inoperable, intravenous chemotherapy was initiated. The cells in the lung tumor responded to the treatment. However, one cell developed a mutation that made it resistant to the chemotherapeutic agent. After a few months, tumors returned to the different parts of Tom's body: the lungs, bones, and liver. Sadly, his body could not take all this, and Tom died of metastatic lung cancer at the age of 76. That is, he died 31 years after the carcinogenic asbestos dust made first contact with his lung.

To summarize, Tom's story shows that cancer develops in several steps. Its multistep development starts with carcinogens triggering genetic mutations.

The mutations disrupt signaling pathways, leading to more mutations and the selection and survival of cancerous cells. In terms of treatment, it is good that cancer develops gradually because, at the beginning of its development, there is time to stop cancer from progressing. A cell has repair mechanisms to correct malfunctions triggered by mutations, or it can destroy itself long before cancer develops. Only when the cell has acquired critical properties does it become a cancer cell, yet this process takes time.

2.8 Immune System

"Imagine that your most dangerous opponent is also your closest friend." This quote from *Modern Cancer* by Hana Heikenwälder perfectly captures the challenge facing cancer research today. Our immune system, while a constant protector against infections and damaged cells, can also become a double-edged sword. It's crucial for preventing the growth of tiny cancer cell clusters and healing wounds [4]. Yet, this same power can sometimes turn destructive.

Typically, we notice our immune system at work only when we're sick. However, it's constantly operating behind the scenes, even when we feel perfectly fine. It's fascinating to consider how often we remain unaware of the immune system's efforts to shield us from potential threats.

To fully appreciate the complexity and efficiency of the immune system, it is helpful to examine what happens after death. Without its protection, the body rapidly falls prey to microbial breakdown by bacteria and fungi. Although these microbes are everywhere, the immune system keeps them in check throughout our lives. Remarkably, it also coexists peacefully with beneficial bacteria, like those in our gut, which aid digestion. These "good" bacteria colonize nearly every surface of our bodies, defending us against harmful invaders. In Chap. 8, we'll dive deeper into this interplay, exploring the microbiome, a vast community of bacteria that outnumbers human cells.

One of the biggest scientific puzzles in recent years is the role the immune system plays in cancer. For a long time, it was considered our most powerful ally in the fight against the disease. However, recent studies suggest that immune responses may also contribute to tumor growth [4]. We now know that in the early stages of cancer, the immune system acts as a defense mechanism. But as tumors grow larger, chronic inflammation can accelerate cancer progression and even aid in the spread of metastases.

In this context, scientists have classified the immune response into "beneficial" and "harmful" components. On the positive side, the immune system's ability to recognize and destroy cancer cells is invaluable. Specialized T-cells, for instance, identify cancer cells by spotting altered surface proteins. This mechanism is also crucial in organ rejection, as the immune system sees foreign tissue as a threat.

However, chronic inflammation—a negative aspect of the immune system—can create an environment that supports tumor growth. In such conditions, some tumors seem to thrive, feeding off the inflammation to grow and spread. Although the exact mechanisms remain unclear, understanding these complex interactions opens up new possibilities for treatment.

Thanks to advances in immunotherapy, we now have new ways to enhance the immune system's ability to fight cancer. Many treatments have been very successful, especially for some types of blood cancer. These include CAR T-cell therapy, which changes the genes of T-cells to target and kill cancer cells, and checkpoint inhibitors, which stop proteins like PD1 and PDL1 from slowing down the immune response. We'll explore these treatments in greater detail in Chap. 6.

But the immune system's importance extends far beyond cancer. It also plays a critical role in vaccinations, allergies, and autoimmune diseases. In autoimmune diseases, the immune system mistakenly attacks healthy tissue, leading to chronic inflammation. Allergies are exaggerated immune responses to harmless substances like pollen or food. Vaccines work by exposing the body to harmless parts of a virus, triggering an immune response that creates a memory for future infections.

Take natural killer cells, for example. These cells are constantly patrolling the body, searching for cells that lack certain "identification proteins," [9] a common trait of aggressive cancer cells. When these proteins are missing, natural killer cells act quickly to eliminate the threat. This surveillance system is key to detecting and destroying tumor cells that evade other immune responses.

The immune system is one of the most fascinating subjects in biology, a complex network of highly specialized cells and signaling molecules. Despite its complexity, it is incredibly adaptable, providing protection in a variety of ways. However, this very flexibility also means that the immune system can sometimes turn against us in specific situations.

Generally, the immune system identifies and eliminates harmful changes before they cause serious problems. The effectiveness of these defense mechanisms explains why most of us never develop cancer, even though our bodies are constantly producing potentially dangerous cells. However, in rare cases where cancer cells outperform all the defense systems, the disease can become life-threatening.

In summary, the immune system is both a defender and, in some cases, a supporter of cancer. Understanding this dual role is critical for advancing cancer treatment and prevention, as well as laying the groundwork for innovative therapeutic strategies.

2.9   Microenvironment

Cancer needs favorable *microenvironment conditions* to grow. Mutated precancerous cells are initially found in an environment with anticancer effects. Unless it is in an inflammatory environment, like Tom's story, they cannot take advantage of the surrounding resources at that point. Additional factors are actually required to make the environment better for cancer. Just as a seed waits for favorable conditions to germinate and mature, cancer does the same. The seed mainly needs adequate sunshine and water to absorb nutrients in the soil. Gardeners know that plants also require a special quality of soil to grow exceptionally well. Similarly, precancerous cells require several factors that enable them to obtain the elements they need for their progression [3] (Fig. 2.15). As Stephen Paget has hypothesized, cancer cells metastasize to other sites because they can thrive better in the "soil" or the environment there. In sum, cancer needs an environment more suited for its survival and growth. Prostate cancer metastases preferentially develop in bone, while intestinal cancer metastasizes to the liver.

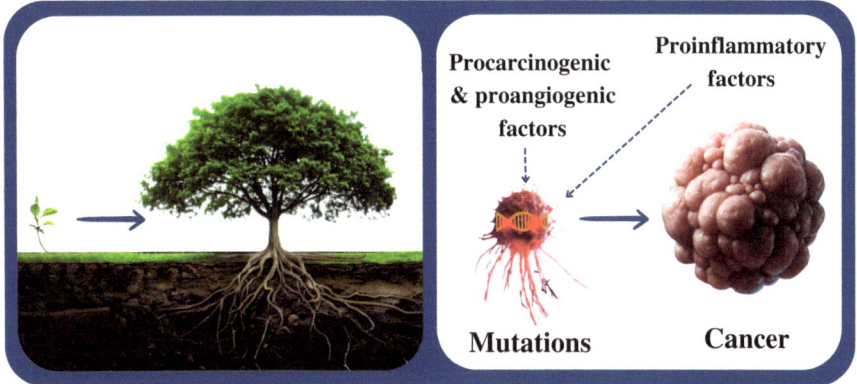

Fig. 2.15 Mutations are like seeds waiting for the right conditions to germinate. Seeds can germinate and grow into a plant when they can access favorable conditions such as sunlight, water, and soil. Similarly, pro-carcinogenic, pro-angiogenic, and pro-inflammatory factors provide an excellent environment for mutations to develop into cancer

2.10 Inflammation

The growth and spread of cancer depend heavily on inflammation. In reaction to injury, infection, or foreign body invasion, the body naturally limits damage and starts the healing process. Still, persistent or uncontrolled inflammation can encourage tumor development. These are fundamental immune system defense reactions. Though the immune response differs depending on the trigger, inflammation always seeks to strengthen the immune system. Immune cells release signaling chemicals in order to attract more immune cells, dilate the blood vessels, and increase blood flow. This leads to symptoms such as redness, swelling, warmth, and pain.

About half of our white blood cells are neutrophils, which cluster in the damaged tissue and combat infections by spewing hostile oxygen molecules (ROS). Large scavenger cells and macrophages eliminate the leftovers and show the immune system the broken-down pathogen bits. Triggered B cells or T cells swiftly proliferate upon recognizing these foreign proteins.

These immune cells pass via lymph fluid to the lymph nodes, which function as gathering places. To guarantee that there are always some cells that identify foreign proteins, the body can theoretically generate over one quadrillion distinct immune cells [16]. This amount, which comprises 200 billion of the 1024 stars in our observable universe, surpasses the total number of stars in our Milky Way by a factor of 5000.

An immune cell quickly multiplies and creates an army to battle a pathogen as soon as it finds one.

The immune system takes around 5 days to create a particular reaction against infections [17]. No matter the treatment, an improvement usually occurs after 1 week. Once we battle the infection, some immune cells remain in the body as memory cells. When a fresh infection strikes, they act quickly to eradicate the organism before disease symptoms appear. Recent research has demonstrated that inflammation promotes cancer growth and spread. Inflammation in specific organs, particularly those involved in inflammatory bowel disease, can directly cause tumors.

Moreover, inflammatory mediators entering the bloodstream can raise the cancer risk in other body organs. In carcinogen-induced malignancies, such as lung cancer caused by cigarette smoke, the inflammatory response is also quite significant. Similar to asbestos, consistent lung irritation from smoke particles increases the risk of lung cancer. This is most likely due to the more than 60 carcinogens in smoke [18]. Almost every malignancy involves an apparent inflammatory reaction at some point. Lack of nutrients causes cells in growing tumors to die. The immune system sets up an inflammatory response, mistaking this cell death for tissue damage. These inflammatory cells help the tumor proliferate and encourage the expansion of the surviving cancer cells. This allows the benign tumor to grow from a benign state to a deadly one, spreading throughout the entire body.

How might inflammation lead to cancer? It is crucial to understand that any other kind of inflammation cannot cause cancer more than a long-term, chronic one. Short-term inflammation that arises from acute infections or temporary damage is insufficient. Chronic inflammation, which is usually invisible for an extended period of time, can be unpleasant or generate no symptoms. Should they not improve or remain untreated, the toxic compounds generated by immune cells can, over time, destroy our bodily cells.

A critical factor in this is reactive oxygen species (ROS), which include hydrogen peroxide (H_2O_2). Neutrophils release these ROS to fight bacteria.

Hydrogen peroxide (H_2O_2) lightens hair due to its bleaching power [4]. It is also quite aggressive at the cellular level. Together with breaking down germs inside neutrophils, these reactive oxygen molecules help to kill microbes.

These drugs can cause significant damage to chronic inflammation, even if they are helpful in acute inflammation. They often interact with proteins found in cells, which causes more complications.

Studies reveal that chronic inflammation—which results from inflammatory bowel disease (e.g., Crohn's disease) or chronic liver inflammation—causes genetic damage capable of fostering cancer. People with inflammatory bowel disease already had genetic modifications in their inflamed intestinal cells before they developed cancer. The same holds for chronic liver inflammation, in which cancer development follows genetic damage to the liver cells.

We expected reactive oxygen molecules to induce DNA mutations directly. We know they initially interact with cell proteins vital for DNA replication and repair. Many mutations in the damaged cells can raise cancer risk.

As they destroy cancer cells, radiation and chemotherapy can aggravate the inflammatory reaction in tumors. This can lead to several outcomes: on the one hand, the immune system can identify the dead cells as a hazard and set off an immunological reaction directed against the tumor [19]. Conversely, if the immune system does not combat the cancer cells successfully, the inflammation may hasten tumor development.

The immune system can assist or stop tumor development, depending on the kind of cancer. Certain cancer types, such as colorectal carcinoma, exhibit a higher chance of survival due to the presence of T cells within the tumor. The absence of these T cells would suggest a shorter life period [20]. Other malignancies, including melanoma and breast cancer, have shown similar trends.

We need to conduct more studies to develop successful immunotherapies and gain a deeper understanding of how the immune system functions in cancer. We remain uncertain about why some tumors disregard the immune system while others benefit from it. Correct control of the immune response is critical since activating or suppressing the immune system could hasten tumor growth or weaken cancer defenses.

Inflammation that lasts a long time can raise the risk of cancer, especially in immune cells. Their deterioration and higher blood cancer risk are consequences of permanent immunological activation. At the same time, Epstein-Barr virus infections can induce lymphomas, including Burkitt's or Hodgkin's lymphomas [3]. Also, infections such as chlamydia or *Helicobacter pylori* raise the risk of MALT lymphomas [21]. Moreover, autoimmune illnesses and persistent inflammation are risk factors for chronic lymphocytic leukemia and potentially multiple myeloma.

Produced during inflammation, reactive oxygen species (ROS) can damage genes and encourage cancer. These substances harm DNA and help cancer grow. Long-term inflammation raises the risk of liver cancer, just as it does in hepatitis B. Hepatitis B immunizations are available. However, the infection is still relatively common, especially in underdeveloped nations. Alcohol consumption and fatty liver disease primarily cause liver cancer in Western countries. Along with inflammation, a fatty liver can turn into either cancer or liver cirrhosis.

A major factor in the development of cancer cells is inflammation. Inflammation occurs as a defensive response to aggression, typically from foreign bodies or microbes. Certain cells, such as fibroblasts, which normally reside at the site of aggression, can produce molecules that facilitate their proliferation to repair injury. These molecules, like fibroblast growth factor, for example, can also stimulate the proliferation of other cells, including cancer cells. There are anticancer treatments known as "fibroblast growth factor inhibitors" that work by blocking this process.

Short Story of the Inflammation Process

When an intruder is detected, inflammatory cells such as the white blood cells release highly reactive molecules to eliminate it. When these reactions occur, the inflammatory cells release growth factors so healthy cells can arrive more quickly, and new blood vessels form more easily in the race to beat cancer.

Hence, "resident" cells are capable of emitting distress signals aimed at summoning defense cells (white blood cells) to aid in combating the aggressor. At the site of this confrontation, symptoms such as irritation, redness, swelling, or itching may manifest.

A strong inflammatory response is now known to accompany practically every advanced cancer, as cells within a rapidly growing tumor eventually become hungry for oxygen and nutrients, leading to their death [22, 23]. The immune system perceives cancer cell death as a tissue injury. In reaction, immune cells employ messenger molecules to increase blood flow inside the tumor and attract more immune cells to repair the damaged area. More and more inflammatory cells penetrate the tumor, prompting the remaining body cells—in this case, cancer cells—to multiply to replace the dead ones.

Additionally, certain specialized white blood cells may release molecules that stimulate the formation of new blood vessels, similar to how an army constructs roads for transporting military resources and personnel. The nutrients and oxygen transported by these new vessels, intended for the defense cells, can inadvertently nourish cancer cells as well. Furthermore, the fragile nature of these newly formed vessels facilitates the infiltration of cancer cells into the bloodstream, a topic we will revisit later. We will also explore the medications developed to counteract the formation of these new vessels, finally, at the site of inflammation, cells [24].

2.11 Cancer Risk Factors

When people are asked about their understanding of what causes cancer, the majority of the respondents think that faulty or mutated genes inherited from our parents are the leading cause. However, **heredity** and genetic influence are not the most important risk factors for cancer to develop. Only rarely can cancer develop due to abnormalities that occur during fetal development or because we inherit faulty or mutated genes from our parents.

Many people also know that environmental factors cause cancer, especially since they blame it significantly on smoking. However, less than half of respondents associate cancer and diets. Migrant Japanese populations have shown that differences in lifestyle, rather than genetics, account for the extreme difference in cancer rates mentioned previously. When the Japanese migrated and adopted a Western diet and lifestyle, the rates of some of their cancer types increased considerably. These data indicate that environment, diet, and lifestyle also have a significant impact on cancer development. Although we will discuss later how these cancer-causing agents and risk

Fig. 2.16 Illustrates the dietary shift from a Western diet to a Japanese diet. Research on migrant Japanese populations indicates that lifestyle changes, rather than genetics, largely account for significant differences in cancer rates

factors lead to cancer development (Fig. 2.16), identifying these factors is crucial to helping us avoid developing the disease.

These findings emphasize the significant role that environmental factors, diet, and lifestyle play in cancer development. Various elements, from sex hormones like estrogen and testosterone to growth and metabolism regulators like insulin and IGF-1, as well as chronic inflammation and physical stimuli, can all contribute to the risk of cancer by hindering the removal of damaged cells. We'll explore how these carcinogens and risk factors contribute to cancer in more detail later, but recognizing them is crucial for disease prevention.

Recent experiments uncovered that most cancer incidences are caused by factors related to our lifestyle and dietary habits (Fig. 2.17). First, *poor dietary habits,* including low fruit and vegetable intake and consuming processed food loaded with sugar and fat, are often linked to cancer. **Second, drinking alcohol** can also cause stomach and intestinal cancer. Third, in some instances,

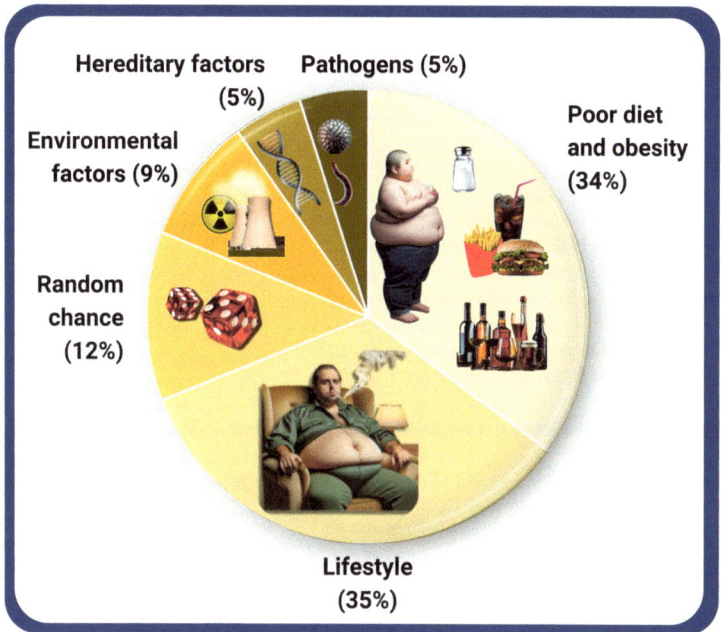

Fig. 2.17 Distribution of cancer risk factors. Several factors play a role in cancer. Lifestyle and diet represent the biggest part (69%) [25]

poor diet leads to ***obesity***, a key factor in cancer development. Fat cells attract the inflammatory cells in the human body; obese or overweight individuals are more prone to developing uterine, esophageal, kidney, colon, and breast cancer [26]. In addition, obesity is also linked with cancer because of the role of altered metabolism of insulin, which is more frequent in obese individuals. Fourth, certain habits, such as a ***sedentary lifestyle,*** also increase the risk of cancer. Indeed, ***physical inactivity*** weakens the body's immune system, creating an environment conducive to cancer cells. Fifth, ***smoking*** tobacco or cigarettes is also responsible for one-third of all cancers. The tar and soot in cigarettes contain known carcinogens, such as nitrosamines and polycyclic aromatic hydrocarbons, which often lead to cancers in the lung, head, neck, stomach, and pancreas [27].

Next, cancer-causing substances also abound in our environment. These ***environmental factors*** lead to genetic mutations and thus increase our risk of cancer. We encounter them every day and are exposed to them through our skin, the air we breathe, the food we eat or the form of radiation we are exposed to. Excessive exposure to natural UV rays from the sun and artificial

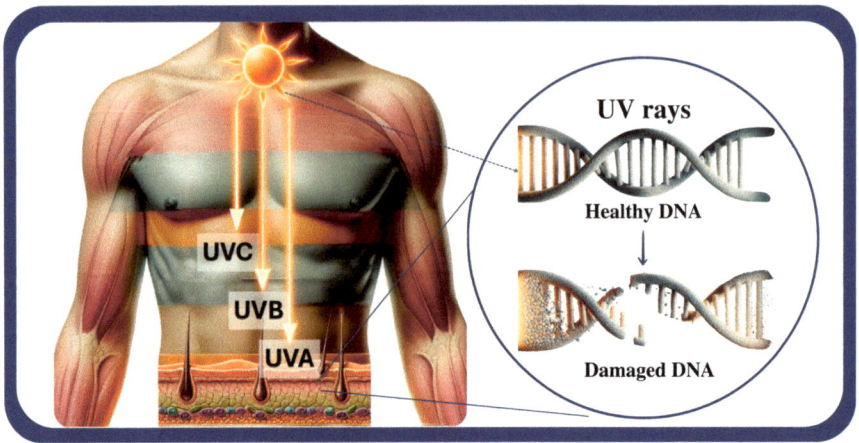

Fig. 2.18 The effect of UV rays on the skin. UV rays damage our healthy DNA, causing skin cancer. UVA rays penetrate deeper into the skin than UVB rays, while UVC rays penetrate our skin the least

sources, such as tanning beds, is associated with skin cancer or melanoma (Fig. 2.18).

Take Australia, for example. That continent has continuously reported high skin cancer rates in recent years. Not far from Australia, the ozone layer, our natural protection against harmful radiation from the sun, has been significantly depleted over Antarctica, leaving us with only a thin protective layer against harmful UV rays. Additionally, Australia is much closer to the sun during summer due to the Earth's orbit, which further exposes Australians to more intense UV rays [28, 29]. History also contributes to the high number of skin cancer cases in Australia. Many modern Australians have a skin type that is not suited to their environment. Most of them originally came from less sunny climates, so their skin lacks protective pigmentation. This makes them more vulnerable to UV rays from the sun, which damages their skin. On the other hand, Aboriginal people in Australia have very low skin cancer rates due to their pigmented skin, which fits their environment [30].

Sometimes, in our jobs, we are exposed to substances that cause cancer, including industrial or naturally occurring hazards (Fig. 2.19). For example, asbestos fiber was formerly used as insulation in buildings. Yet, we now know that asbestos frequently causes lung and throat cancer. In addition, sawdust in wood processing is the leading cause of malignant cancer in the sinus. Ionizing radiation caused by radon gas can lead to skin cancer. Chemical carcinogens, such as benzene in gasoline and smoke, can damage our genes. Vinyl chloride used to make PVC is believed to cause liver cancer. Heavy metals such as

Fig. 2.19 Substances in the environment that cause cancer. Our environment contains many hazardous substances, such as asbestos fiber, sawdust, radiation, benzene, vinyl chloride, and heavy metals. Exposure to these substances increases cancer risk

mercury, lead, chromium, cadmium, arsenic, and nickel can damage cells and have a carcinogenic effect as they cause DNA damage. These metals often enter our bodies via contaminated water or food, such as fish, seafood, rice, or chocolate. Mercury, which mainly accumulates in predatory fish such as tuna, is a neurotoxin and can cause chemical burns but is not a carcinogen. Heavy metals such as cadmium, chromium, arsenic, and nickel, which often come from industrial sources, can cause nasal and lung cancer. Usually, we recognize these risks only when their effects become evident [31]. Looking back, there are many examples where people were unknowingly exposed to very harmful carcinogens. For instance, in the past, small young boys were hired as chimney sweeps. They climbed inside chimneys to clean them. These boys regularly worked naked, exposing themselves to toxins in the chimney soot, which

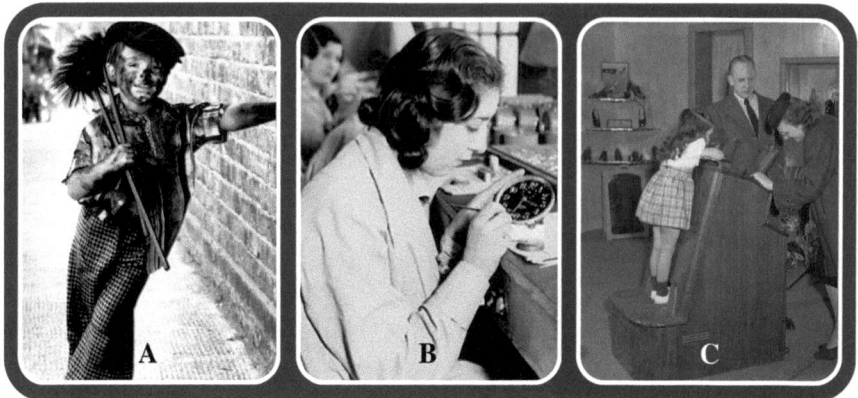

Fig. 2.20 Cancer in the past. (a) Young boys hired as chimney sweeps developed testicular cancer in their later years. (b) Watch painters who frequently licked brushes with radium-infused paint suffer from radium toxicity. (c) Shoe-fitting fluoroscopes utilizing X-ray radiation were previously used in shoe shops, which increased rates of foot cancer

would stick to their sweaty skin. Decades later, they began to develop scrotum warts, which later progressed to scrotal cancer, a form of testicular cancer [32] (Fig. 2.20a).

In the 1910s, after the discovery of radium, the US Radium Company began to create a radium-infused paint that emitted a greenish-white light in the dark. This paint was used on clock dials to create glow-in-the-dark watches. The watch painters frequently licked their brushes to produce sharp lines. Then, they began complaining of jaw pain, fatigue, and skin and tooth problems. Without knowing it, they were suffering from the effects of radium toxicity, and many soon died of leukemia and other cancers (Fig. 2.20b).

Finally, from the 1920s through the 1950s, shoe stores used shoe-fitting fluoroscopes with X-rays to see if the shoes fit before the customer bought them. But these machines were stopped as they were suspected of causing foot cancer in salesmen and in children who were often exposed to that radiation [33] (Fig. 2.20c).

Pathogens can also trigger cancer. These pathogens include viruses, bacteria, and parasites. Oncogenic DNA viruses change the host cell's DNA so that proto-oncogenes, tumor suppressor genes, and the genes that regulate cell death can no longer function correctly.

Impaired immunity, including HIV infection, leads to several different cancers, including non-Hodgkin's lymphoma and HPV-associated malignancies, such as anal and cervical cancer [27]. Their unique shapes allow for precise interactions, either with other proteins or within themselves. These

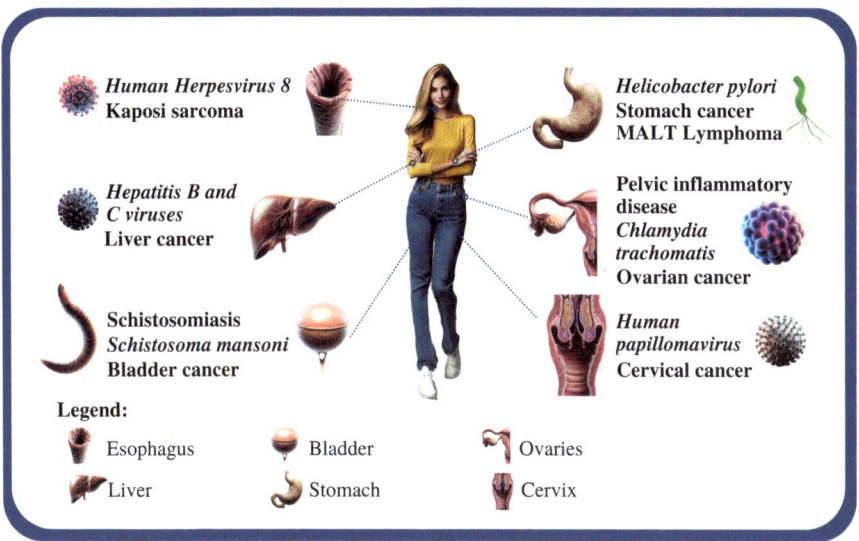

Fig. 2.21 Infectious diseases that increase the odds of getting cancer. Infection and inflammatory diseases caused by numerous foreign substances and pathogens can affect organs and develop into cancer

interactions can convey important messages within the cell, such as signals for proliferation. The sequences of interlocking and shape-shifting events passed from one protein to another to transmit these messages are known as signaling cascades or pathways. Certain bacteria like *Helicobacter pylori* also affect organs, increasing the risk of getting cancer or MALT lymphomas [21] (Fig. 2.21).

In contrast to Australia, where skin cancer prevails, in Asia, a continent boasting 4.2 billion inhabitants, equating to 55% of the global population, chronic infections like viral hepatitis emerge as the primary cancer risk factor [34]. In China, while lung, breast, and colon cancers are the most frequently diagnosed, lung, liver, and stomach cancers exhibit the highest mortality rates, likely mirroring the nation's rapid industrialization and shifting socioeconomic dynamics. Chronic infections such as *Hepatitis B* virus, *Hepatitis C* virus, and *Helicobacter pylori*, along with smoking, exert significant influence. Gastric cancer remains a prominent issue in Asia due to higher rates of *H. pylori* infection, specific dietary habits, smoking, genetic mutations in the MALT gene, and increased alcohol consumption [35, 36]. Understanding these factors is crucial for improving patient care. For example, the defective *CagA* gene causes DNA damage, which the body typically repairs with the help of genes like *BRCA1* and *BRCA2* through a process called homologous recombination. However, if these genes mutate and lose their functionality, the repair process is hindered, thereby increasing the risk of cancer [37].

In addition to environmental factors and infections, hormones such as estrogen play a crucial role in cancer development, particularly in breast tissue. Estrogen, primarily a female sex hormone, is a key factor in promoting breast cancer. Hormonal contraceptives and hormone replacement therapies that contain estrogen can increase the risk of breast cancer, especially after menopause. Being overweight further heightens this risk, as fatty tissue contributes to estrogen production. Estrogen activates the estrogen receptor, promoting cell growth in the breast and uterus. However, its full cancer-promoting effect occurs mainly in individuals with certain genetic predispositions, such as those with *BRCA1* or *BRCA2* mutations. All body cells contain a critical DNA repair enzyme. However, under the influence of estrogen, these mutations particularly increase the risk of cancer in the breast and uterus.

An estimated 1 million of the 20 million breast cancer cases diagnosed each year are due to hormone replacement therapy after menopause. Taking oral contraceptives increases breast cancer risk by 24%, and this elevated risk only returns to normal about 10 years after stopping the pill [38]. Being overweight raises the risk by 90%, as adipose tissue significantly boosts estrogen production (along with the lower production in the adrenal cortex). Typically, women who develop breast cancer after menopause have 15% higher estrogen levels in their blood compared to healthy women [39].

Generally, a woman's risk of breast cancer decreases with fewer menstrual cycles over her lifetime. Factors like late onset of menstruation, frequent pregnancies, extended breastfeeding, and early menopause are all associated with a significantly lower risk of breast cancer. The age of first menstruation has steadily decreased over the past century, while the onset of menopause has experienced a delay. Currently, the average age for first menstruation is 12 years, with menopause beginning around 51.4 years. For every 1-year delay in menopause, the relative risk of breast cancer increases by 2.9% [38].

In some instances, getting cancer may also be due to chance. Random triggers are those against which we cannot actively protect ourselves. Such random triggers include genetic errors in cell division. The body's toxins also play an influential role in cancer. The body produces its own by-products, such as hydrogen peroxide, that can cause harmful oxygen radicals. As we saw in Chap. 1, free radicals are toxins that create mutations that can lead to cancer.

Lastly, the final cancer risk factor we discuss here is *age* [4]. Age is one of the most important risk factors. In fact, in 2017, only 13% of cancer deaths occurred in people under 50 [40]. Why is age an important factor? As we get older, our exposure to carcinogens from the environment increases. The damage these toxins cause builds up over time, and with age, our immune system weakens. When our immune system is old, it fails to target those precancerous cells. Nevertheless, even though cancer is extremely prevalent among the elderly (e.g., prostate cancer affects almost one in two men over the age of 80) [41], this does not necessarily imply that this illness will be the reason for their death. As we saw earlier, cancer cells require time to grow and manifest themselves. Numerous elderly individuals may experience additional health issues that may arise before the extent of their cancer exposure.

Aging was once seen as an unavoidable process of accumulating cellular and tissue damage. Today, we comprehend the close connection between aging and cancer. Rare genetic variants found in exceptionally long-lived people suggest that it is possible to influence the aging process, thereby reducing the cancer risk. For instance, a 70-year-old man is 1000 times more likely to develop bowel cancer than a 10-year-old boy [4]. In 90-year-old patients, bowel tumors exhibit nearly twice as many genetic mutations as those in 45-year-olds [5]. These findings indicate that cancer is largely a matter of time—if you live long enough, it becomes more likely. In fact, autopsies reveal that 60–70% of deceased individuals had tumors that went undetected during their lifetime [39].

The shortening of telomeres, the protective caps on our chromosomes, would still limit our lifespan. Among the very elderly, around the age of 110, blood stem cells become depleted, increasing susceptibility to infections and thus shortening life expectancy [4, 42]. The frequency of blood stem cell division over a lifetime, such as from chronic inflammation or infections, influences this depletion. Frequent cell division not only increases the risk of DNA damage and blood cancer but also shortens telomeres, resulting in a shorter lifespan.

A troubling trend in antiaging medicine involves giving older people blood transfusions from young donors to enhance health and appearance. However, growth hormones from young donors could introduce unpredictable cancer risks. Additionally, lifestyle habits like smoking, obesity, and regular alcohol consumption can also shorten telomere length and, consequently, lifespan. On the bright side, a healthy lifestyle can positively influence telomere length without increasing cancer risk. Practices such as a calorie-restricted diet, regular exercise, adequate sleep, and effective stress management help protect telomeres and reduce cancer risk. Nobel Prize winner Elizabeth Blackburn advocates these methods for cancer prevention.

Studies of very old individuals reveal that they often carry genetic mutations in genes critical for nutrient perception and cellular stress response. Centenarians frequently carry mutations in the IGF-1 receptor gene [43], crucial for cell growth, and the insulin receptor gene, which significantly contributes to sugar metabolism. These gene mutations, including those in *FOXO1*, *FOXO3a*, and *AKT* [4], are common among long-lived people worldwide and may provide an evolutionary advantage by mitigating the harmful effects of excess food. This could suggest that high sugar insulin levels may shorten the lifespan of both humans and other mammals.

All the above information on risk factors suggests we can sometimes weaken cancer or prevent its development by removing the favorable conditions that help it thrive in the first place. For example, eating certain types of cancer-fighting food can create a hostile environment that causes the microtumors to regress. As we saw earlier, microtumors in Japan are a prime example of this. These tumors have a poorer prognosis when Japanese patients live and eat in Japan compared to countries where Western food is consumed. We can also change our odds of getting cancer by making different lifestyle choices. By changing our bad habits and adopting good ones, we can stop or minimize our exposure to external factors that can lead to cancer. Avoidance of risk factors reduces the risk of developing cancer, but it does not completely prevent its development. The good news today is that cancer has begun shifting from a fatal disease to a chronic illness that can be controlled through constant treatment and better awareness.

2.12 The Hallmarks of Cancer

By this point, we can see that cancer is a very complex disease. Cancer cells are mainly defined by their uncontrolled cell division. But that is not all. Metastasis occurs when cancer cells migrate to different body parts and invade other organs. That makes treatment harder. Cancer scientists thus want to completely understand cancer's characteristics and identify how mutations make cells behave abnormally. They also want to know why healthy cells sometimes turn into malignant cancer cells. Cancer cells behave a little like baby cells, which divide and grow quickly, and at the same time like blood cells, which can move anywhere in the body.

Cancer has many characteristics. This is why it has defied our common logic, and its rules seem to be nonexistent. Healthy cells have physiological limits, which they respect, but tumor cells transcend these boundaries. For a good cell to become a cancer cell, it must undergo specific mutations and damage the systems that usually work together to prevent cancer. Just as a hijacker can take control of a plane and turn it into a flying bomb, cancer hijacks normal biochemical processes for its own destructive reasons. Thanks to research, scientists have uncovered the mechanisms behind consistent behaviors, the genes, and signaling pathways. The signaling pathways facilitate their development and explain how malignant transformations take place, leading to the spread of cancer or its escape to other organs. These traits are present in all malignant cancer cells, although they were not created by cancer cells. Rather, they came from distorting the body's normal processes. This happens because cancer cells bypass the cell cycle rules. That said, what exactly are the hallmarks, features, or characteristics of cancer? Recently, 14 hallmarks of cancer have been reported [44]. However, we will discuss just the ten most important ones here.

The first and most significant characteristic of cancer cells is their **unrestrained cell division.** As we saw earlier, oncogenes are activated in cancer cells and interfere with signaling regulation. The most common mutated oncogene in tumors is the *RAS gene. RAS gene* mutations allow cells to divide independently and become self-sufficient.

The second important hallmark of cancer is that its cells can sustain **proliferative signaling.** Oncogene mutations are only one of the many reasons why cancer cells can do this. Cancer can interfere with normal signaling regulation to sustain proliferative signaling instead. Cancer cells possess the ability to stimulate their own proliferation internally through the signaling pathways, which we have previously discussed through analogy with messages, planes, and airports. They are also capable of secreting external messages or molecules that facilitate their own proliferation. They can send themselves messages to proliferate. Consider a situation where rabbits multiply uncontrollably. When rabbits were brought to Australia, there were just a few of them. However, because they multiply very fast, they have become an invasive species, causing major problems for the country's agriculture and croplands (Fig. 2.22). Just as rabbits in Australia proliferated without much control, cancer cells can do the same thing in our body. This is why chemotherapy and radiotherapy try to treat cancer by targeting proliferating cancer cells.

Third, cancer cells can also **evade growth suppressors** and checkpoints. They can do that by inactivating tumor suppressor genes that usually stop the cell cycle at the correct times. Tumor suppressor genes are the guardians of the cell cycle. They help to slow down cell division by giving the cell enough time to do all the checks and balances during division. Just like us, if we don't have enough time and have to do many things quickly, we might do them wrong and make mistakes. A cell that divides too quickly without a plan can turn into a different type of cell. But when they are inactivated or suppressed, cancer cells become insensitive to signals that try to inhibit their growth. That enables cancer cells to grow and proliferate uncontrollably despite the cellular

Fig. 2.22 Rabbit invasion in Australia. Only a few rabbits were introduced to Australia at first, but since then, they have become an invasive species. They multiply at a fast rate and produce large numbers of offspring. Similarly, cancer cells can sustain proliferative signaling and unrestrained cell division. As a result, they quickly grow in number and can wreak havoc on the body

stress and DNA damage they cause. To see how this works, imagine a bank guarded by two security officers. If both are incapacitated, then the thief can rob the bank (Fig. 2.23). The two tumor suppressor genes are like those two bank guards; cancer is the robber, and our body is the money that cancer wants. Today, this cancer hallmark is targeted by using *CDK4/6* inhibitors.

The fourth hallmark involves ***controlled cell death.*** Typically, when healthy cells sense DNA is damaged, they initiate programmed cell death or controlled cell death, also known as ***apoptosis***. However, cancer cells can suppress and inactivate genes and seize the signaling pathways that can cause apoptosis. When cancer cells hijack these pathways, they become resistant to controlled cell death. Controlling cell death is like a computer's ability to detect errors, such as computer viruses. When the computer fails to detect a virus, it can stop working correctly. However, it might also shut itself down to prevent further damage (Fig. 2.24). This defensive move can occur thanks to antivirus software that detects and eradicates bad programs before they can cause damage. Similarly, cancer cells have their own antivirus software. They can thus resist controlled cell death by over-activating anti-apoptotic molecules such as *BCL-2*. In lymphoma patients, a *BCL-2* inhibitor called venetoclax is already used to target this hallmark.

Fig. 2.23 If the bank's security officers are out of action, robbers can easily infiltrate. Likewise, cancer cells can grow uncontrollably by inactivating both tumor suppressor genes

The fifth hallmark is ***the life expectancy of a cell***. In normal cases, healthy cells divide and replicate around ***40 times***. Then they die of old age, unlike stem cells, which can continually regenerate themselves throughout life. However, by activating specific signaling pathways in genes, cancer cells can keep on replicating and dividing without any limits. This means they have replicating immortality, even after many generations. It is like the Marvel superhero Wolverine, who becomes immortal. He has an accelerated healing factor that makes him immortal. As cancer cells can replicate endlessly, they also seem immortal.

Telomerase plays an essential role in cell immortality. Telomeres are found at the ends of the chromosomes, which act as special caps that shield the ends of DNA sequences from damage. Telomeres are just like aglets at the ends of shoelaces that prevent them from fraying (Fig. 2.25).

Telomeres become shorter with each cell division and wear down with aging. When the telomeres are used up, the cell dies, just as a shoelace eventually snaps. They shorten with each cell division and wear out with aging.

Fig. 2.24 Error message. An error message pops up to protect the computer from further damage. In the worst cases, an error caused by a virus also forces the computer to shut down

When telomeres are exhausted, the cell dies, just as when the end of a shoelace is so worn that it no longer protects anything (Fig. 2.25). Imagine a mutation that prevents the lace ends from wearing out: they'll always be protective, and the laces will never wear out. Mutations that prevent telomere wear prevent cell death. Cancer cells become immortal by constantly activating telomerase, which replenishes the telomeres as they dwindle (Fig. 2.26). An analogy can be made with the branches of the starfish and the tail of the lizard, which can "grow back" when cut. The same type of regeneration—albeit endless in the case of cancer—occurs in cancer cells. Drugs targeting telomerase activities are currently under development.

In addition to telomerase, mechanisms like the Shelterin protein complex and the protein POT1 help regulate telomere length [45]. Researchers at Johns Hopkins University have found that individuals with mutations in the *POT1* gene had longer telomeres and a more youthful appearance but were at higher risk for tumors and cancer. This indicates that while longer telomeres can delay aging, they also increase the likelihood of cancer due to the accumulation of mutations over a lifetime.

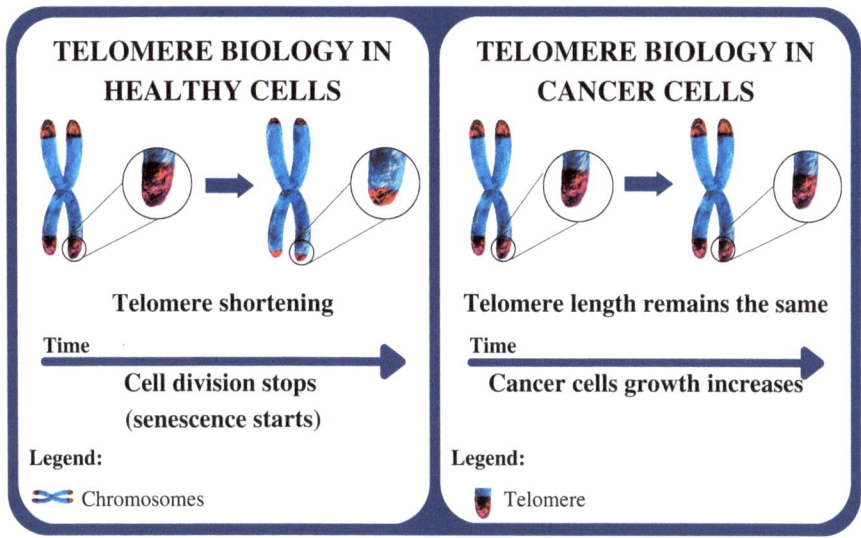

Fig. 2.25 This figure illustrates how the telomere at the end of a chromosome functions like the aglet on a shoelace. Just as the aglet protects the end of a shoelace from fraying with use, the telomere protects the end of a chromosome during cell division. However, just as an aglet wears out over time with repeated use, telomeres progressively shorten with each cell division, ultimately leading to cell aging and dysfunction

Understanding telomeres and their role in cell aging and division is crucial, as it could lead to advancements in our knowledge of aging and the development of new cancer therapies. Currently, there are drugs in development that target telomerase activity, offering potential new treatments for cancer.

The sixth hallmark is ***angiogenesis***. This occurs when cancer cells hijack the body's balance, causing new blood vessels to form and feed cancer. Tumor cells release specific messenger molecules that encourage the growth of new blood vessels in their surrounding tissue, which are about two millimeters in size at first. These micro-blood vessels are like fiery volcanic fissure cracks (Fig. 2.27). Tumor tissue needs plenty of nutrients and oxygen because of its high proliferative power. Hence, tumors try to form new blood vessels a lot. As these new vessels are immature and show cracks, they, therefore, bleed frequently, which facilitates metastasis. The cancer cells can thus travel to other parts of the body. Today, many treatments, such as Avastin, target how microvessels are built. This is another strategy in cancer treatment.

A seventh hallmark of cancer is ***metastasis***. This is the ability of cancer cells to escape from their primary site, migrate, and spread to other organs in the body. In other words, they activate ***invasion and metastasis***. They loosen from their surrounding tissue, reach and penetrate the blood vessels, pass

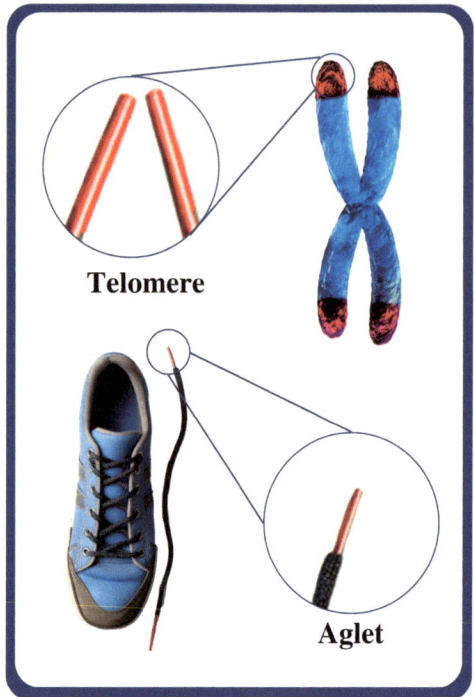

Fig. 2.26 Telomeres and cancer cells. Telomeres shorten in healthy cells, but cancer cells permanently activate telomerase to keep the length of the telomeres the same, causing cancer cells to become immortal

through the blood or the lymphatic system, and float in the blood until they can exit at another location, where they can colonize another tissue. Just as soldiers destroy the defenses of a castle they want to invade, cancer cells invade and colonize healthy tissues (Fig. 2.28). Treatments before (neoadjuvant) or after surgery (adjuvant) help destroy metastatic cells while they are still invisible and vulnerable before they grow and become indestructible.

The eighth hallmark of cancer is that cancer cells can avoid destruction by the immune system. In other words, they escape the immune system or avoid being detected by it. Typically, when the immune system detects cancer cells, it proactively eliminates most of them. Unfortunately, it cannot eliminate all of them. The cancer cells that were left behind reach equilibrium, where the immune system controls the growth of the cancer tumor and the tumor becomes resistant to destruction. When the immune system cannot detect cancer cells, they are free to continue growing and dividing. But how do cancer cells go undetected? To visualize this, imagine people in a masquerade ball; they wear masks that hide their real faces. However, as everyone is wearing a

Fig. 2.27 Volcanoes spew magma from their cracks and veins. Similarly, due to unbalanced angiogenesis in leaky blood vessels, cancer cells leak out from them, and metastasis occurs

mask, a thief could easily blend in with the crowd and not be caught (Fig. 2.29). One type of cancer therapy, called immune checkpoint inhibition, unmasks cancer cell proteins by blocking them from interacting with healthy immune cells.

The ninth hallmark of cancer cells is that they have a ***heightened sensitivity to mutagens,*** which cause genetic changes in DNA. These changes can promote ***genome instability*** and increase ***mutation*** rates. That makes it favorable for mutations to accumulate and sustain tumor growth. In other words, when DNA changes, it can help cancer develop.

Finally, the tenth hallmark of cancer is that it ***promotes inflammation***. Tissue inflammation helps cancer spread by supplying it with growth and survival factors. These factors accelerate cancer formation. Strong anti-inflammatory treatments, such as cortisone, are frequently used in oncology, often providing temporary relief by reducing the inflammatory content of the tumor. Cancer is also known to ***deregulate cellular energetics***. Cancer cells reprogram energy production so it supports their uncontrolled proliferation. They reprogram energy by adjusting glucose metabolism, upregulating

Fig. 2.28 Cancer cells are like invaders. They destroy the body's defenses and invade and kill healthy tissues. As a result, metastatic growth starts

glucose transporters and using other alternate pathways. To obtain enough energy, cancer cells reabsorb a higher amount of glucose than a normal, healthy cell would. This reabsorption ability is a unique feature of cancer cells and is used for diagnostic purposes, which we will explain later in this book.

Cancer treatment therapies can target these subtle differences in genes, signaling pathways and acquired capabilities that make cancer cells different from normal cells. The idea is to shoot an arrow into the cancer cells' Achilles' heel without damaging the healthy cells. This is challenging, as both healthy and cancerous cells come from the same organism, which means they share many similarities. We can make treatments for this problem by noticing the small differences between them.

To sum up, whether a tumor is benign or malignant, it contains errors or changes known as mutations. These mutations can occur in important DNA segments, and driver mutations disrupt the cell's normal functions. On the other hand, passenger mutations mainly cause copying errors that remain insignificant and are just passed on.

Fig. 2.29 A masked thief remains undetected at a masked ball where everyone else is also wearing a mask. He can realize his evil intentions by blending in with the crowd

There are four classes of genes that are altered in tumor cells. Members of these four classes are also called key genes. The first key gene altered in cancer cells is the ***proto-oncogene***; it activates proliferation and cell growth. In normal cells, they are activated mostly during repair processes, such as after an injury, but also when the person needs to build up an "army" of white blood cells quickly to defend against infection. If a proto-oncogene is altered, this gene causes cell division to become unregulated and uncontrolled. Thus, at this point, the proto-oncogene becomes a cancer gene called an ***oncogene***. The second key gene is the ***tumor suppressor gene***; each cell has two of them, and they usually stop uncontrolled cell growth and suppress tumor development. If the two copies of this gene are inactivated, normal functioning is lost, and the cells start dividing nonstop. It is like when a car loses its brakes and drives very fast without any control. A third key gene is the ***repair gene***. Under ordinary conditions, these genes are utilized to create the apparatuses necessary to rectify DNA mutations. These occur during cell division or when the cell encounters a substance that causes a mutation. If both copies of the repair genes are defective, the apparatuses can no longer be produced, and DNA errors cannot be corrected. If the repair gene is broken, it allows cancer to continue. The ***apoptotic gene*** is the fourth and final key gene that plays a crucial role in maintaining the health and functionality of cells. When a cell becomes so badly damaged that it poses a threat to the organ it belongs to, the apoptotic gene triggers a process known as "cell suicide." This mechanism prevents further damage and allows the organ to continue functioning properly. In other words, the apoptotic gene is a vital component of the body's defense mechanisms, ensuring its survival and well-being. If they are lost, cancer formation occurs. When cells malfunction, it could stop important signaling pathways from working normally or abnormally turn on other unwanted signaling pathways. These disruptions can cause cancer cells to grow uncontrollably, escape death, and survive in hostile environments. This is why these factors are often targeted when drugs are developed to prevent cancer cells from growing.

Certain factors cause mutations to occur. Contrary to widespread beliefs, mutations passed on through heredity comprise only a small percentage of factors that cause cancer. Most mutations that cause cancer come from poor dietary habits and certain activities, such as smoking. In our daily lives, we also encounter substances in the environment that cause cancer. Cancer can also be caused by pathogens that cause infectious diseases. Last but not least, a good number of mutations also occur by chance, randomly.

All these cancer-risk factors contribute to mutations, which reveal why there is so much heterogeneity in the disease. However, regardless of the factor that led a patient to develop cancer, all cancer cells behave consistently. They have specific characteristics and capabilities summarized as the Hallmarks of Cancer. When the cell's normal processes are disrupted, the reason may be cancer. Doctors who understand the characteristics of cancer and cancer cells can screen patients for the disease and diagnose it, even at its early stages. The next chapter will discuss this process.

References

1. Masters JR. HeLa cells 50 years on: the good, the bad and the ugly. Nat Rev Cancer. 2002;2(4):315–9. https://doi.org/10.1038/nrc775.
2. Callaway E. Deal done over HeLa cell line. Nature. 2013;500(7461):132–3. https://doi.org/10.1038/500132a.
3. Ferreri AJM, Ernberg I, Copie-Bergman C. Infectious agents and lymphoma development: molecular and clinical aspects. J Intern Med. 2009;265(4):421–38. https://doi.org/10.1111/j.1365-2796.2009.02083.x.
4. Heikenwälder H, Heikenwälder M. Der moderne Krebs—Lifestyle und Umweltfaktoren als Risiko. Berlin/Heidelberg: Springer Berlin Heidelberg; 2023. https://doi.org/10.1007/978-3-662-66576-3.

5. Vogelstein B, Papadopoulos N, Velculescu VE, Zhou S, Diaz LA, Kinzler KW. Cancer genome landscapes. Science. 2013;339(6127) https://doi.org/10.1126/science.1235122.

6. Chaffer CL, Weinberg RA. A perspective on cancer cell metastasis. Science. 2011;331(6024):1559–64. https://doi.org/10.1126/science.1203543.

7. Wie entstehen Antibiotikaresistenzen? https://www.bag.admin.ch/bag/de/home/krankheiten/infektionskrankheiten-bekaempfen/antibiotikaresistenzen/wie-entstehen-antibiotikaresistenzen%2D%2D-.html. Accessed 05 July 2024.

8. Holtkamp W. Cancer. German ed. Springer Berlin Heidelberg

9. Der König aller Krankheiten: Krebs—eine Biografie: Mukherjee, Siddhartha, Pleitgen, Fritz, Schaden, Barbara: Amazon.de: Books. https://www.amazon.de/K%C3%B6nig-aller-Krankheiten-Krebs-Biografie/dp/3832196447. Accessed 05 July 2024.

10. Aaronson SA. Oncogenes and the Molecular Origins of Cancer. Robert A. Weinberg, Ed. Cold Spring Harbor Laboratory Press, Cold Spring Harbor, NY, 1989. Xii, 367 pp., Illus. Paper, $55. Cold Spring Harbor Monograph 18. Science. 1990;249(4973):1177–8. https://doi.org/10.1126/science.249.4973.1177-a.

11. Definition of signaling pathway—NCI Dictionary of Cancer Terms—NCI. https://www.cancer.gov/publications/dictionaries/cancer-terms/def/signaling-pathway. Accessed 05 July 2024.

12. Jordan JD, Landau EM, Iyengar R. Signaling networks. Cell. 2000;103(2):193–200. https://doi.org/10.1016/s0092-8674(00)00112-4.

13. Folkman J, Kalluri R. Cancer without disease. Nature. 2004;427:787. https://doi.org/10.1038/427787a.

14. Food as Medicine—Dr. William Li at Exponential Medicine. 2020. https://www.youtube.com/watch?v=qhJZcKFfu_c. Accessed 05 July 2024.

15. Informationen zu Krebs | dkfz—Krebsinformationsdienst. https://www.krebsinformationsdienst.de/. Accessed 05 July 2024.

16. Lythe G, Callard RE, Hoare RL, Molina-París C. How many TCR clonotypes does a body maintain? J Theor Biol. 2016;389:214–24. https://doi.org/10.1016/j.jtbi.2015.10.016.

17. Murphy K, Weaver C. Janeway immunologie. Berlin/Heidelberg: Springer Berlin Heidelberg; 2018. https://doi.org/10.1007/978-3-662-56004-4.

18. Takahashi H, Ogata H, Nishigaki R, Broide DH, Karin M. Tobacco smoke promotes lung tumorigenesis by triggering IKKbeta- and JNK1-dependent inflammation. Cancer Cell. 2010;17(1):89–97. https://doi.org/10.1016/j.ccr.2009.12.008.

19. Zitvogel L, Apetoh L, Ghiringhelli F, Kroemer G. Immunological aspects of cancer chemotherapy. Nat Rev Immunol. 2008;8(1):59–73. https://doi.org/10.1038/nri2216.

20. Galon J, Costes A, Sanchez-Cabo F, Kirilovsky A, Mlecnik B, Lagorce-Pagès C, Tosolini M, Camus M, Berger A, Wind P, Zinzindohoué F, Bruneval P, Cugnenc

P-H, Trajanoski Z, Fridman W-H, Pagès F. Type, density, and location of immune cells within human colorectal tumors predict clinical outcome. Science. 2006;313(5795):1960–4. https://doi.org/10.1126/science.1129139.

21. Vela V, Juskevicius D, Dirnhofer S, Menter T, Tzankov A. Mutational landscape of marginal zone B-cell lymphomas of various origin: organotypic alterations and diagnostic potential for assignment of organ origin. Virchows Arch. 2022:480. https://doi.org/10.1007/s00428-021-03186-3.

22. Mantovani A, Allavena P, Sica A, Balkwill F. Cancer-related inflammation. Nature. 2008;454(7203):436–44. https://doi.org/10.1038/nature07205.

23. Karin M. Nuclear factor-κB in cancer development and progression. Nature. 2006;441(7092):431–6. https://doi.org/10.1038/nature04870.

24. Beliveau R. Foods to fight cancer: essential foods to help prevent cancer. 1st ed. New York: DK; 2007.

25. AACR_CPR_2011.Pdf. https://cancerprogressreport.aacr.org/wp-content/uploads/sites/2/2020/09/AACR_CPR_2011.pdf. Accessed 05 July 2024.

26. Westerlind KC. Physical activity and cancer prevention—mechanisms. Med Sci Sports Exerc. 2003;35(11):1834–40. https://doi.org/10.1249/01.Mss.0000093619.37805.B7.

27. What Causes Cancer? https://www.news-medical.net/health/What-Causes-Cancer.aspx. Accessed 05 July 2024.

28. Anthony Augustine. Why is skin cancer so common in Australia? SunDoctors. https://sundoctors.com.au/blog/why-skin-cancer-is-common-australia/. Accessed 05 July 2024.

29. Specialist Clinics of Australia. Why is the skin cancer rate higher in Australia? Specialist Clinics of Australia. https://specialistaustralia.com.au/why-is-the-skin-cancer-rate-higher-in-australia/. Accessed 05 July 2024.

30. Why does Australia have so much skin cancer? (Hint: it's not because of an ozone hole)—Cancer Council WA. https://cancerwa.asn.au/news/why-does-australia-have-so-much-skin-cancer-hint-i/. Accessed 05 July 2024.

31. BVL2018—Startseite—Seite nicht gefunden. https://www.bvl.bund.de/DE/Arbeitsbereiche/01_Lebensmittel/02_. Accessed 30 Aug 2024.

32. Blogging the Human Genome: why so many London chimney sweeps suffered scrotal cancer. https://slate.com/technology/2012/07/blogging-the-human-genome-why-so-many-london-chimney-sweeps-suffered-scrotal-cancer.html. Accessed 05 July 2024.

33. The era of the shoe-fitting fluoroscope and the radiation it caused. https://interestingengineering.com/health/the-era-of-the-shoe-fitting-fluoroscope-and-the-radiation-it-caused. Accessed 05 July 2024.

34. Southern, Eastern, & South-Eastern Asia. The Cancer Atlas. http://canceratlas.cancer.org/mD0. Accessed 10 July 2024.

35. Shin WS, et al. Updated epidemiology of gastric cancer in Asia: decreased incidence but still a big challenge. Cancers Basel. 2023;15(9) https://doi.org/10.3390/cancers15092639.

36. Huang J, et al. Cancer incidence and mortality in Asian countries: a trend analysis. Cancer Control. 2022;29:10732748221095955. https://doi.org/10.1177/10732748221095955.

37. Ärzteblatt, D. Ä. G., Redaktion Deutsches. Neue Erkenntnisse zu Helicobacter-pylori-Resistenzen. Deutsches Ärzteblatt. https://www.aerzteblatt.de/nachrichten/145773/Neue-Erkenntnisse-zu-Helicobacter-pylori-Resistenzen. Accessed 04 July 2024.

38. Britt KL, Cuzick J, Phillips K-A. Key steps for effective breast cancer prevention. Nat Rev Cancer. 2020;20(8):417–36. https://doi.org/10.1038/s41568-020-0266-x.

39. Weinberg RA. The biology of cancer. 2nd ed. New York: W.W. Norton & Company; 2013. https://doi.org/10.1201/9780429258794.

40. Dettmer P, Vogel S, Flückiger A. Immun: Alles über das faszinierende System, das uns am Leben hält. Das Immunsystem erklärt vom Macher des beliebten.

41. Jahn JL, et al. The high prevalence of undiagnosed prostate cancer at autopsy: implications for epidemiology and treatment in the PSA era. Int J Cancer. 2015;137(12):2795–802. https://doi.org/10.1002/ijc.29408.

42. Walter, et al. Reported that exiting dormancy triggers DNA-damage-induced attrition in hematopoietic stem cells. Nature. 2015;520(7548):549–52.

43. Kenyon CJ. The genetics of ageing. Nature. 2010;464(7288):504–12. https://doi.org/10.1038/nature08980.

44. New dimensions in cancer biology: updated hallmarks of cancer published—American Association for Cancer Research (AACR). https://www.aacr.org/blog/2022/01/21/new-dimensions-in-cancer-biology-updated-hallmarks-of-cancer-published/. Accessed 12 July 2024.

45. DeBoy EA, et al. Familial clonal Hematopoiesis in a long telomere syndrome. N Engl J Med. 2023;388(26):2422–33. https://doi.org/10.1056/NEJMoa2300503.

3

Cancer Screening

Contents

Abstract Cancer screening plays a crucial role in early detection and prevention, often saving lives by identifying cancers before symptoms appear. Common cancers that benefit from screening include breast cancer, cervical cancer, colorectal cancer, and prostate cancer, which can often be detected in early stages through various screening methods. This chapter explores various screening methods, such as mammograms, Pap smears, colonoscopies, and prostate exams, which are commonly used to detect precancerous conditions and early-stage cancers. These tests are particularly important for high-risk individuals, and national organizations like the Swiss Cancer League focus on the most common cancers to maximize resources. However, not all cancers are screened for due to rarity or limitations in test accuracy. Patients can still discuss personalized screening options with their doctors. The chapters highlight recent technologies, including artificial intelligence (AI), that in recent years has begun transforming cancer diagnostics. AI enhances screening by increasing the accuracy of detecting diseases like breast cancer in mammograms and improving colonoscopy results by identifying adenomas. Prostate cancer screening is also benefiting from advancements like the PSA test and the 4K score, which help reduce unnecessary procedures. As AI continues to evolve, its potential to revolutionize cancer detection and treatment grows, offering a promising future for proactive healthcare strategies aimed at reducing the global cancer burden.

3.1   Screening

Prevention is better than a cure. In the age of big data, for healthcare systems worldwide, it has become a priority to get ahead of cancer and stop it before it even begins. Screening tests have many benefits, including detecting diseases such as diabetic eye disease, detecting anomalies in unborn babies during pregnancy, detecting any life-altering conditions in a newborn baby, and detecting many other types of cancer. Yet, screening tests are not meant to diagnose cancer. In fact, screening someone for cancer means checking to see if that person is at risk of developing cancer and then checking to see if any cancer may be present without our knowledge. Screening tests are not offered to the entire population due to factors such as cost, effectiveness, and the need to focus on individuals at higher risk. They are also only offered when a treatment is available that, if used early, would lead to a cure or change the prognosis of the disease. It is inappropriate to offer a test that detects metastatic cancer for which there is no evidence that early treatment will improve the prognosis.

This clarifies why we only offer screening tests when we deem the risk of a disease to be significant. In cases where early treatment doesn't affect the course of the disease, such examinations are not useful. People at increased risk are the focus of examinations for cancer screening. It wouldn't be cost-effective to screen the entire population if the risk of developing cancer is very low. For instance, various countries only recommend mammography screening for breast cancer for women between the ages of 40 and 50.

Being screened for cancer is like going through a checkpoint at the airport before boarding an airplane (Fig. 3.1). Just as passengers are screened for deadly weapons to ensure the flight is safe, screening people for cancer aims to reduce the number of lives taken by cancer and ensure their safety. Catching cancer before it can make its move, therefore, makes sense. However, cancer screening is carried out selectively on people at risk of developing cancer. It would be too expensive and inefficient to test the whole population when the risk of getting cancer is very low. For example, women should have a

Fig. 3.1 Cancer screening is like passing through an airport security checkpoint. Both are done to identify and catch dangerous items that can compromise one's safety and endanger life

mammogram to check for breast cancer between the ages of 40 and 50, depending on where they live. It's not a good idea to have a mammogram for young women because they don't usually get this type of cancer. Early screening is only recommended in cases of family risk.

Screening involves surveying the population broadly and then focusing on individuals with an above-average risk of developing the disease. For the general population, the average serves as a reference value. Concentrating on at-risk groups allows doctors to pay more attention to those who are particularly vulnerable. This approach is preventive rather than reactive. Through screening, doctors can save thousands of lives each year by detecting cancer before it fully develops, for example, by removing precancerous polyps. Early diagnosis and appropriate treatment can save lives. It's important to note that this is an active approach to cancer diagnosis that targets the general population, including healthy individuals. It is a massive challenge to effectively and accurately screen an entire nation for cancer, requiring various resources and well-organized systems. Regular screening is essential to prevent the omission of unexpected developments. Unfortunately, individual healthcare facilities can not handle this task alone. Through national healthcare systems, the government often takes responsibility for cancer prevention. In Switzerland, organizations such as "Swiss Cancer Screening" and the "Cancer League" aim to provide all citizens with access to high-quality cancer screening. The Swiss cantons have set up and implemented screening programs, routinely inviting people over 50 to participate. Additionally, these health organizations promote the establishment of new programs and collaboration between existing initiatives.

Screening first looks at an entire population. Then, the focus is narrowed down to those whose risk of developing a disease is higher than normal. What is normal is the benchmark established by an entire population. This narrowing down process enables doctors to pay special attention to the individuals most at risk. This is a proactive strategy rather than a merely reactive one. Thanks to the information that screening provides doctors, thousands of lives are saved every year by catching cancers early and sometimes even before they become cancerous. For example, intestine polyps are removed before they become cancerous. By detecting cancers at early stages and by treating patients with the right therapy, doctors save lives every year.

In the future, artificial intelligence will also play a significant role. AI is revolutionizing cancer diagnosis and treatment, allowing for the deciphering of complex cancer patterns and improving early detection. AI-assisted cancer screening has the potential to enhance diagnosis and treatment and ultimately save lives. We will discuss three more detailed examples of the interaction of artificial intelligence in cancer diagnostics.

3.2   Artificial Intelligence

Artificial intelligence (AI) is transforming cancer screening and diagnosis by identifying high-risk individuals with increased accuracy. Techniques such as deep learning, convolutional neural networks (CNNs), and machine learning algorithms analyze imaging data to detect anomalies with greater precision, reducing false positives and improving early detection rates. For example, a research group in Chicago and Boston has developed a deep learning model called "CXR-Lung-Risk," which assesses lung cancer risk in non-smokers using chest X-rays. Although smoking is the primary cause of lung cancer, some cases occur in individuals who have never smoked. Traditional screening methods often overlook such individuals, but AI-driven models enhance detection by analyzing large datasets. Current screening recommendations focus mainly on smokers, and there are few reliable ways to predict the risk of lung cancer in kidney smokers. The CXR-Lung-Risk model was trained using a large dataset of over 147,000 chest X-rays, including asymptomatic and kidney smokers. Once developed, the model was validated on a separate group of kidney smokers. It requires only a single X-ray to assess the risk of lung cancer, making it a feasible method for clinical practice. The results of the model are promising as it is able to identify kidney smokers with an increased risk of lung cancer. This could lead to improved early detection and treatment of the disease in this specific population. The study highlights the potential of AI and deep learning in medicine, particularly in identifying individuals at increased risk of lung cancer who have never or only lightly smoked [1]. By analyzing X-ray images from electronic medical records, the CXR-Lung-Risk model was able to provide accurate results. It offers a promising method for determining the risk of lung cancer in people who have never smoked. This development could have a positive impact on the design of screening programs and the early detection of lung cancer, especially in people whose risk is difficult to assess due to their smoking habits.

In Stockholm, a risk model based on artificial intelligence [2] has been developed to identify women with an increased risk of breast cancer based on mammography images. The AI-based model takes family history and lifestyle factors into account and recognizes patterns with minimal changes to the images. It considers thousands of factors to provide an estimate of future breast cancer risk. The study confirmed that the AI model was able to identify a group of women who had a breast cancer risk almost seven times higher than the average population. The researchers think adapted screening would be more appropriate for this high-risk group.

In another study on mammography in Munich, the impact of artificial intelligence on the detection rate of cancer was investigated. In 2022, the Transpara AI system was used to analyze 28,000 mammograms and compared with the 30,000 mammograms from the previous year without the use of AI [3]. The AI determined a risk score for each mammogram and classified the cancer risk as low, medium, or high. The study found that the cancer detection rate increased significantly from 4.4 to 5.8 per 1000 women through the use of AI. The researchers concluded that the use of AI increases the cancer detection rate and enables improved risk assessment.

These studies show that AI-based systems are not only able to detect breast cancer more effectively but also assess the risk of individual patients more accurately. This could lead to a personalized screening approach, which would be particularly beneficial for women at increased risk.

It is important to note that this is a proactive approach to a cancer diagnosis, and it is made at a general population level, which includes healthy people. Of course, it is challenging to screen the population of an entire country for cancer. Such a task requires a huge number of people and systems to be in place to make it happen. Such a task also requires an advanced level of organization to ensure the whole process runs smoothly so that individuals at risk are identified accurately. Furthermore, screening must be done regularly to ensure nothing unexpected pops up in the meantime. As we can see, screening can be complex.

Unfortunately, no single healthcare facility can handle this big burden alone. This is where the government steps in. A national-level healthcare system often undertakes screening for cancers. Consulting a doctor or a specialist, like a family doctor or oncologist, is another way to find people who are "especially at risk" because of their genes, jobs, or smoking habits. These people may be examined outside of national screening programs. In the following section, we will present another study for you.

In Cambridge, Massachusetts, and Boston, researchers at MIT are developing an AI named "OncoNPC" [4] to assist oncologists in identifying the primary tumor in patients initially diagnosed with only metastases, a condition known as CUP syndrome. This AI was trained using a dataset from over 36,000 patients from three large cancer clinics. It analyzed the genetic profiles of the tumors and compared them with known cancers. When tested on 971 patients with CUP syndrome, the AI successfully identified the tumor of origin in 400 cases (41.2%). It also discovered specific gene signatures associated with certain types of cancer, such as mutations in the *EGFR* gene in lung cancer or mutations in the *PIK3CA* gene and changes in the *CCND1* gene in breast cancer. In pancreatic cancer, the AI detected mutations in the *KRAS* gene, which are crucial for this tumor.

3.3 Mammography

Breast cancer screening is done with periodic breast X-rays called ***mammograms***. This test involves using ***low-dose X-rays*** to take images of each breast. With the help of a special X-ray machine, each breast will be flattened between two plastic plates to hold it in place while images are taken [5]. The whole mammogram process usually takes about 15–30 min from start to finish.

Thanks to mammography, 80% of malignant tumors are detected. Consequently, many deaths from breast cancer are prevented. The 20% of tumors that are not detected by mammography may be detected during the subsequent examination 2 years later if their evolution is slow. They may also be detected by palpation by the patient or her gynecologist. It's important to remember that mammography can also show benign abnormalities, and as we'll see later, only by getting rid of the abnormality can cancer be officially diagnosed. In fact, mammography helps doctors detect various breast tumors and cysts even before the patient may feel tumors present. In just 5–10% of breast cancer cases, genetic predispositions (such as mutations in the *BRCA* gene) are the cause. In women with such mutations, it is recommended to perform more precise tests, such as breast MRI.

New studies indicate that PET & CT diagnostics (explained in Chap. 4) can be a valuable method in the treatment of patients with HER2-positive breast cancer. These examinations make it possible to dispense with chemotherapy in some patients without compromising survival. In a study in Barcelona, 356 patients with HER2-positive breast cancer were treated with antibodies. PET & CT images were used to decide whether additional chemotherapy was necessary for some of the patients [6]. If the images showed a good response to the antibody therapy, chemotherapy was not administered. Many patients benefited from this strategy and showed excellent survival without invasive disease. The results of this study are encouraging and indicate that PET & CT diagnostics could play an important role in the individualized treatment of breast cancer. The most common risk factors for breast cancer are age and hormonal factors. For these reasons, women over a certain age are encouraged to have mammograms every 2 years.

- The German surgeon Albert Salomon performed the first mammography study in 1913.
- In 1937, the American researcher and physician Stafford L. Warren developed a form of mammography to diagnose breast cancer at earlier stages.
- American radiologist Philip Strax led the 1966 study, which pioneered the widespread clinical use of mammography as a screening technique.

3.4 Pap Smears

Cervical cancer screening is done by routine ***cervical/pap smears*** in women. Cervical cancer can develop due to a permanent infection from the human papillomavirus, which is passed on during unprotected sexual intercourse.

The cervical/pap smear scans the cervix for cell abnormalities that could develop into cancer if left untreated. During this procedure, the gynecologist will examine the vagina and the cervix. Then, cells and mucus from the cervix and surrounding areas are extracted with a cotton swab. The cells collected are sent to a laboratory and are checked for any pathological changes [7]. While the examination itself only takes a few minutes, results usually come back from the laboratory within a few weeks. Women are advised to undergo this test every two years. After the age of 65–70, the need for further tests should be discussed with the gynecologist, depending on the results of previous tests and the risk of contracting the papilloma virus, i.e., sexual activity.

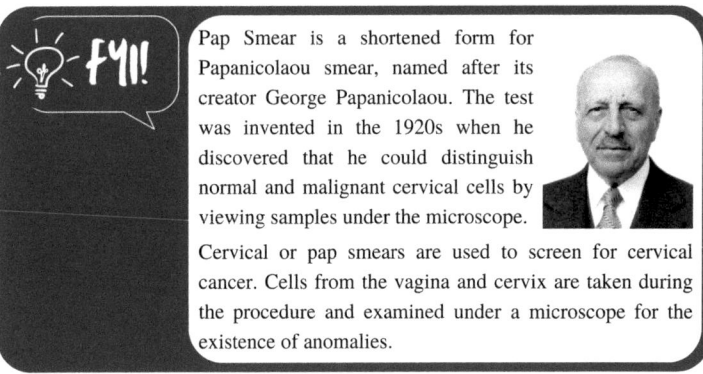

FYI!

Pap Smear is a shortened form for Papanicolaou smear, named after its creator George Papanicolaou. The test was invented in the 1920s when he discovered that he could distinguish normal and malignant cervical cells by viewing samples under the microscope.

Cervical or pap smears are used to screen for cervical cancer. Cells from the vagina and cervix are taken during the procedure and examined under a microscope for the existence of anomalies.

3.5   Screening the Colon

Colorectal cancer screening occurs in two ways. The first method involves periodic *colonoscopies* or *sigmoidoscopies* in adults, usually after 50 years of age. In this type of examination, the inside of the intestine is inspected with a tiny camera located on the *endoscope* that is inserted into the intestine (Fig. 3.2). Then, the images of the intestine are projected on the screen in real time. During a colonoscopy, abnormal tissues such as polyps (growths of tissue on the colon's lining) can be removed when necessary. Tissue samples can also be taken during a colonoscopy. This technique enables doctors to directly remove tumors during the colonoscopy. There is a high level of certainty in detecting tumors with this method.

Thus, we can prevent cancer from developing thanks to colonoscopies. The entire process usually takes between 30 and 60 min and is often performed under anesthetic. If everything appears normal and no tissue samples are taken for further analysis, then colonoscopy findings can be retrieved immediately.

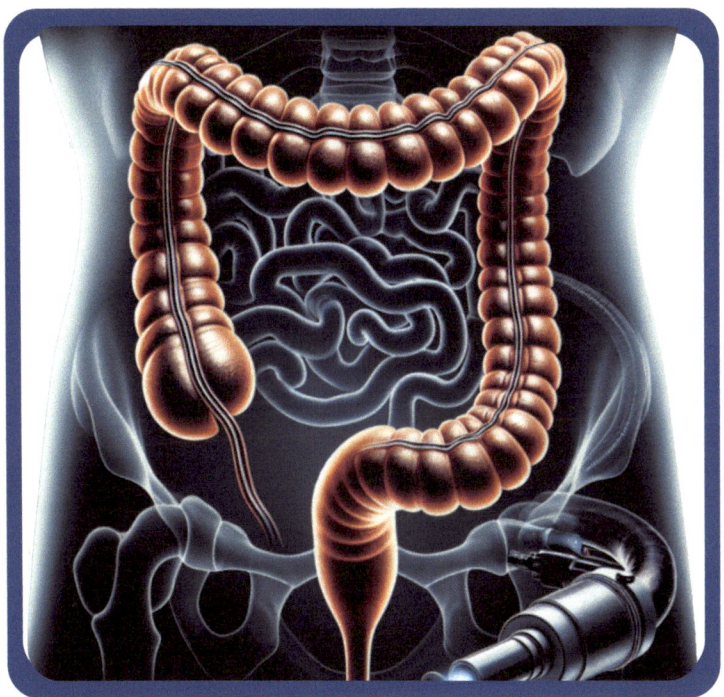

Fig. 3.2 Colon cancer is screened using colonoscopy. In the procedure, a tiny camera at the end of an endoscope is inserted into the large intestine via the anus. The presence of abnormal cells is identified through the images projected on the screen

The second method of colorectal cancer screening is a stool test to check for blood. Called the ***Fecal Occult Blood Test,*** it is also used to screen patients for colorectal cancer. If it detects a little invisible blood, it will serve as the basis for requesting a colonoscopy because this may be the first sign of an intestinal tumor. Blood that could be the result of a polyp or cancer is invisible to the naked eye. Unlike the "red" blood that can be seen on stools that are already fully formed when you have hemorrhoids, for example, a drop of blood mixed with stool or that has lost its red color because it has been partially "digested" may not be visible when you look at the stool.

Dr. William Wolff and Dr. Hiromi Shinya, from Beth Israel Medical Center in New York City, created the first fiberoptic colonoscope in 1969. Doctors could inspect the colon's full-length thanks to this equipment. At the same time, Dr. Shinya also invented the polypectomy snare, a device used to remove colorectal polyps physically.

However, it may take a few days or weeks to obtain accurate results. If these fecal blood tests are done ***annually,*** three out of 1000 people may be prevented from dying from colorectal cancer [8].

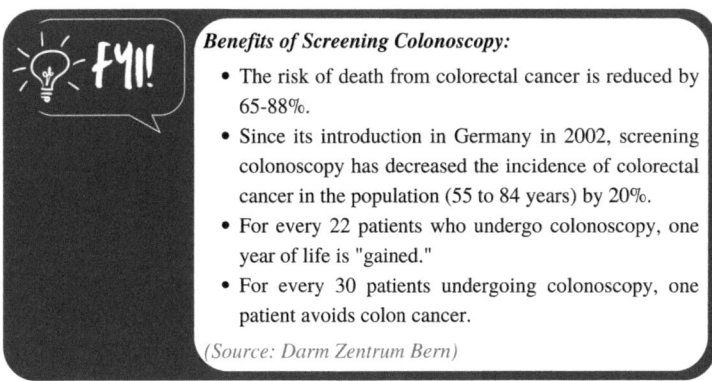

Benefits of Screening Colonoscopy:
- The risk of death from colorectal cancer is reduced by 65-88%.
- Since its introduction in Germany in 2002, screening colonoscopy has decreased the incidence of colorectal cancer in the population (55 to 84 years) by 20%.
- For every 22 patients who undergo colonoscopy, one year of life is "gained."
- For every 30 patients undergoing colonoscopy, one patient avoids colon cancer.

(Source: Darm Zentrum Bern)

New studies published in the New England Journal of Medicine [9] show promising new approaches to testing stool samples. Researchers from Indianapolis and Boston have presented two noninvasive tests based on stool or blood samples that have the potential to improve colorectal cancer screening. The BLUE-C study evaluated a stool test that analyzes molecular DNA markers and hemoglobin content in stool. This test was shown to be sensitive and could detect colorectal carcinomas and advanced lesions in the study involving 20,000 participants.

The ECLIPSE study presented a blood test that examines cell-free DNA in the blood. This test also showed remarkable sensitivity in the detection of colorectal cancer and could play a crucial role in early detection in the future.

Additionally, a study in "Lancet Digital Health" [10] has demonstrated that using computer-assisted detection (CAD) in combination with artificial intelligence can significantly reduce the number of adenomas missed during colonoscopies. CAD-assisted colonoscopy flags suspicious lesions in real time and has proven to be more effective than the conventional method in a study of 900 patients. The results show that CAD colonoscopy detected 37% more adenomas than the traditional method and missed 17% fewer adenomas. The study suggests that CAD technology could provide valuable support in the early detection of colorectal cancer.

These innovative approaches demonstrate that colorectal cancer screening is continuously advancing. The combination of different methods and technologies can help improve survival rates and effectively combat the spread of colorectal cancer.

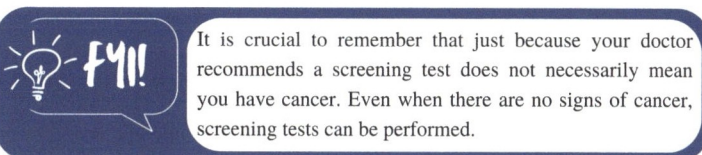

It is crucial to remember that just because your doctor recommends a screening test does not necessarily mean you have cancer. Even when there are no signs of cancer, screening tests can be performed.

3.6 Prostate Cancer Screening

Detecting prostate cancer early is crucial because it significantly improves the chances of treating it at a more manageable stage. Men aged 50 and over should consider regular screening [11]. While most cases of prostate cancer go undetected throughout a person's life, some types grow rapidly, spread aggressively, and metastasize to nearby organs. To identify these more dangerous forms of cancer in their early stages—when they are still confined to the prostate and more treatable—screening tests are essential. Prostate cancer screening typically involves two procedures.

The first test is the ***prostate-specific antigen (PSA) test***. PSA is a protein formed in the prostate that enters the blood in small quantities. A ***blood sample*** is taken and sent to a laboratory for analysis. People with prostate cancer often have elevated blood levels of PSA, which we will discuss later.

Another test for prostate cancer that is often completed right after the PSA test is the ***digital rectal examination (DRE)***. This test involves inserting fingers into the rectum and feeling the prostate to check for abnormalities. The examination will take about 1–2 min. It is usually painless but may cause some discomfort.

In Sweden, a study showed that by combining an elevated PSA level with a blood test called the 4K score [12], fewer biopsies are needed to find prostate cancer. This not only reduces the physical and psychological burden on patients but also the healthcare costs associated with unnecessary biopsies. PSA screening followed by biopsies can reduce the risk of death from prostate cancer. However, the PSA level is not very specific, leading to many unnecessary biopsies. The 4K score, a test that takes into account the age of the patient to better predict whether someone has cancer or not, improves the accuracy of the PSA screening test. The combination of these methods offers a better way of detecting prostate cancer than using the PSA test alone. MRI and the 4K score can help reduce the number of unnecessary biopsies while ensuring that dangerous cancers are not missed. Researchers recommend the 4K score as a way to find out who really needs a biopsy and thus avoid unnecessary examinations, potentially revolutionizing the way prostate cancer is diagnosed and managed.

In a significant development, researchers in Stuttgart [13] have discovered a potential biomarker in the blood that could transform pancreatic cancer detection. This biomarker, based on the methylation pattern of cell-free DNA in blood plasma, shows promise in distinguishing pancreatic cancer from benign diseases such as pancreatitis. If further validated, this could be a significant step toward early detection of these cancers, potentially improving patient outcomes and survival rates.

Apart from the recommended screening examinations, additional tests are organized based on individual and family risk factors, as discussed earlier, for patients without specific complaints. Of course, in cases of symptoms, tests are organized based on the patient's complaints, as we will explore later, but this does not constitute screening.

Here are a few examples of patients at higher risk of developing certain cancers who are monitored more closely "without screening": in a patient who has undergone surgery for testicular cancer, the contralateral testicle will be monitored more closely by ultrasound; in a patient operated on for an abnormal mole, regular dermatological surveillance is advised, with each suspicious mole examined under a special magnifying glass; in a patient with family members affected by ovarian cancer, genetic testing is encouraged, and regular ovarian examinations may be considered. These tests are not carried out for everyone either because certain cancers are too rare to warrant them for the general population or because the tests are not precise or powerful enough to detect them.

Screening tests can be done for many cancers not limited to the above, such as ovarian, skin, pancreatic, and testicular cancers. However, at the level of a national organization like the "Cancer League," only those cancers mentioned above are regularly screened for and are proven to have life-extending benefits. You, as an individual, can still discuss with your doctor the benefit of a cancer screening if you wish, which is then done on an individual basis instead of on a national level.

3.7   Vaccination

Vaccination exhibits an effective prevention against the occurrence of diseases caused by different pathogens including bacteria and viruses. The goal of a vaccine is to train our body to recognize harmless versions of pathogens and build an immune memory. If we are later exposed to the same pathogen, our immune system will quickly identify and eliminate it. The immunity achieved through bacterial and viral vaccinations over the past decades has gained global importance for cancer prevention. However, developing vaccines for cancer is much more difficult because, unlike bacteria and viruses, which are foreign invaders, cancer cells arise from our own body and closely resemble healthy cells. Additionally, cancer develops ways to evade the immune system weakening the immune response, and cancer cells are unique, with specific antigens. These complexities make the creation of effective cancer vaccines particularly challenging [14].

Effective cancer-preventive vaccines are those targeting cancers associated with viral infections. For example, vaccination programs against human papillomavirus (HPV) have been available since 2006 and are recommended as a preventive measure for both females and males starting at age 11 with the aim to reduce cervical cancer, head and neck squamous cell carcinoma (HNSCC), and anal cancer. HPV vaccines are highly effective in adolescents and provide long-lasting protection in adulthood. A second preventive cancer vaccine is the one targeting hepatitis B virus (HBV), available since early 1980, which prevents the development of HBV-related liver cancer. The vaccine is recommended by the World Health Organization (WHO) for infants soon after birth [15].

Yet, the majority of tumors have unknown or absent infection association and no preventative vaccine for nonviral-associated cancers has been approved. As previously mentioned, the main concerns of developing a cancer vaccine are the fear of harming healthy tissue, due to the close similarity in cancer and healthy cells in our body. In this way, cancer vaccines have shifted their focus as therapeutic vaccines, which will be explained in Chap. 6.

However, there's been a concerning rise in other cancers, such as bowel, breast, and prostate cancer, not just in affluent regions but also in developing countries. This trend highlights the urgent need to more effectively apply current scientific knowledge and preventive measures to fully harness the potential for cancer prevention [16]. With the progress of new technologies, including next-generation sequencing (NGS) and artificial intelligence, new hope for developing preventive cancer vaccines is emerging.

According to the International Agency for Research on Cancer, applying existing knowledge could prevent half of all cancers [17]. This entails avoiding carcinogenic factors and taking advantage of available prevention and diagnostic options. For instance, vaccinations against hepatitis B and HPV could prevent around one million cancer cases annually. Infections contribute to 18% of all cancers, while diet is responsible for 35%, and environmental pollution and radiation together account for 7% [18].

Summary

Screening tests can be done for many cancers, not only those mentioned above. For example, there are tests for ovarian, skin, pancreatic, and testicular cancers, too. These tests are not done for everybody, either because certain cancers are too rare to implement on the whole population or because the tests are not precise or powerful enough to detect cancers. However, a national organization like the Krebsliga tends to focus on those cancers mentioned above, as they are the most common. Patients nationwide are thus regularly screened for them, and screening has been proven to save or extend lives. As an individual, however, you can still discuss with your doctor the benefits of a cancer screening. In this case, it is done on an individual basis.

Cancer prevention is increasingly important globally. The goal is to detect cancer early and prevent its onset. Screening tests, such as mammograms, Pap smears, and colonoscopies, are crucial as they can identify precancerous conditions and early stages of cancer. These tests are regularly conducted nationwide and have been proven to save or extend lives. However, screening tests are offered only when effective treatment options exist to reduce disease risk or improve prognosis, focusing resources on individuals at increased risk.

Conversely, individuals can consult their doctor about the benefits of cancer screening on a personal basis. Artificial intelligence (AI) is revolutionizing cancer diagnostics by enhancing early detection and personalized treatment. For instance, AI can improve breast cancer detection in mammograms and increase the detection of adenomas in colonoscopies. For prostate cancer, PSA testing and digital rectal examination are utilized, and the 4K score can reduce unnecessary biopsies. National healthcare systems and organizations like the Swiss Cancer League advocate for access to high-quality cancer screening and the establishment of screening programs. Regular screening is critical to avoid overlooking unexpected developments. However, the challenge of screening an entire nation demands extensive resources and well-organized systems. As technology evolves, the future of cancer screening promises greater efficiency, accuracy, and accessibility, ultimately reducing the global cancer burden. Innovations such as AI-driven diagnostics, liquid biopsy techniques, and advanced imaging technologies are paving the way for more precise, non-invasive, and personalized screening approaches.

References

1. RSNA 2024 Scientific Assembly & Annual Meeting | RSNA. https://www.rsna.org/annual-meeting. Accessed 06 July 2024

2. Eriksson M, et al. European validation of an image-derived AI-based short-term risk model for individualized breast cancer screening—a nested case-control study. Lancet Reg Health—Eur. 2024;37 https://doi.org/10.1016/j.lanepe.2023.100798.

3. Ärzteblatt, D. Ä. G., Redaktion Deutsches. KI-Unterstützung kann Krebserkennung im Mammografiescreening verbessern. Deutsches Ärzteblatt. https://www.aerzteblatt.de/nachrichten/144293/KI-Unterstuetzung-kann-Krebserkennung-im-Mammografiescreening-verbessern. Accessed 04 July 2024.

4. Moon I, et al. Machine learning for genetics-based classification and treatment response prediction in cancer of unknown primary. Nat Med. 2023;29(8):2057–67. https://doi.org/10.1038/s41591-023-02482-6.

5. CDC. Screening for breast cancer. Breast Cancer. https://www.cdc.gov/breast-cancer/screening/index.html. Accessed 05 July 2024.

6. Pérez-García JM, et al. 3-year invasive disease-free survival with chemotherapy de-escalation using an 18F-FDG-PET-based, pathological complete response-adapted strategy in HER2-positive early breast cancer (PHERGain): a randomised, open-label, phase 2 trial. Lancet. 2024;403(10437):1649–59. https://doi.org/10.1016/S0140-6736(24)00054-0.

7. Screening for Cervical Cancer | Cervical Cancer | CDC. https://www.cdc.gov/cervical-cancer/screening/index.html. Accessed 05 July 2024.

8. Hewitson P, Glasziou P, Irwig L, Towler B, Watson E. Screening for colorectal cancer using the faecal occult blood test, hemoccult. Cochrane Database Syst Rev. 2007;2011(1):CD001216. https://doi.org/10.1002/14651858.CD001216.pub2.

9. Chung DC, Gray DM 2nd, Singh H, et al. A cell-free DNA blood-based test for colorectal cancer screening. N Engl J Med. 2024;390(11):973–83. https://doi.org/10.1056/NEJMoa2304714.

10. Maas MHJ, Neumann H, Shirin H, et al. A computer-aided polyp detection system in screening and surveillance colonoscopy: an international, multicentre, randomised, tandem trial. Lancet Digit Health. 2024;6(3):e157–65. https://doi.org/10.1016/S2589-7500(23)00242-X.

11. Prostatakrebs Diagnostik Urologie. Kantonsspital St.Gallen. https://www.kssg.ch/urologie/leistungsangebot/prostatakrebs-diagnostik-und-prostatabiopsie. Accessed 04 July 2024.

12. Ärzteblatt DÄG Redaktion Deutsches. Thema prostatakarzinom. Deutsches Ärzteblatt. May 29, 2024. https://www.aerzteblatt.de/nachrichten/151731/Prostatakrebsscreening-Blutbiomarker-erspart-Biopsie-MRT-und-Ueberdiagnosen. Accessed 04 July 2024.

13. Hartwig C, Müller J, Klett H, et al. Discrimination of pancreato-biliary cancer and pancreatitis patients by non-invasive liquid biopsy. Mol Cancer. 2024;23(1):28. https://doi.org/10.1186/s12943-024-01943-x.
14. Sellars MC, Wu CJ, Fritsch EF. Cancer vaccines: building a bridge over troubled waters. Cell. 2022;185(15):2770–88. https://doi.org/10.1016/j.cell.2022.06.035.
15. Grimmett E, Al-Share B, Alkassab MB, et al. Cancer vaccines: past, present and future; a review article. Discov Oncol. 2022;13(1):31. https://doi.org/10.1007/s12672-022-00491-4.
16. Soerjomataram I, Bray F. Planning for tomorrow: global cancer incidence and the role of prevention 2020–2070. Nat Rev Clin Oncol. 2021;18(10):663–72. https://doi.org/10.1038/s41571-021-00514-z.
17. Tran KB, Lang JJ, Compton K, et al. The global burden of cancer attributable to risk factors, 2010–19: a systematic analysis for the Global Burden of Disease Study 2019. Lancet. 2022;400(10352):563–91. https://doi.org/10.1016/S0140-6736(22)01438-6.
18. Aggarwal BB, Vijayalekshmi RV, Sung B. Targeting inflammatory pathways for prevention and therapy of cancer: short-term friend, long-term foe. Clin Cancer Res Off J Am Assoc Cancer Res. 2009;15(2):425–30. https://doi.org/10.1158/1078-0432.CCR-08-0149.

4

Diagnosing Cancer

Contents

Abstract Now that we know what cancer is and how cancer cells behave, the next question on your mind may be about diagnoses. A patient who is suspected of having a cancerous tumor may have an array of questions to ask, such as: "What type of cancer do I have?" "Is my cancer dangerous?" "What is the exact diagnosis?" "Where is my cancer located?" "Has my cancer spread?" "Is my cancer treatable?" To answer these questions, doctors must first precisely diagnose cancer. Only then can they choose the correct treatment with optimal results at a reasonable cost while avoiding as many side effects as possible. As we can see, doctors must take time to understand the disease as much as possible.

As discussed in previous chapters, each cancer in each person is unique. It is also true that no single diagnostic test can firmly determine the presence of cancer except histology, which is when we study the microanatomy of the cells, tissues, and organs with a microscope. A complete evaluation is thus essential, and it begins by physically examining patients and obtaining their thorough medical history. This is why we need patients to take part in the diagnostic process.

But how do medical professionals determine whether someone has cancer? What tools and techniques do they use in their pursuit of a diagnosis? How does each diagnostic tool offer us crucial information on where the cancer is, what it is, and what can be done about it? As we saw in the previous chapter, there are many ways to screen patients for cancer, and as we will see in this chapter, there are various ways to diagnose cancer. Recognizing that there is a problem is the first step toward solving any issue. For cancer, diagnostic testing involves examinations and techniques to first determine if there is cancer. Not every lump means cancer. But if the lump is a cancer tumor, then the tumor type, its extent, and location, as well as its grade and developmental stage, have to be determined. This is what it means to make a diagnosis. It is important to note that not all lung cancers are treated the same way. Some can be operated on; others cannot. Some will be sensitive to certain treatments, as we saw earlier, and others won't. Some individuals will receive chemotherapy, while others will not.

In this chapter, we will explore some popular methods medical staff use to ascertain that very possibility. As we shall see throughout this chapter, the range of available techniques is wide. The choice of which ones to use or to combine is the doctor's choice to make. It depends on the symptoms, the previous exam results and the availability of the exams. However, it is good to know for ourselves what those methods are and how they help the diagnostic process. We will start with generic techniques and then move on to specialized methods and tests.

4.1   Medical History

The first step is to identify the patient's symptoms and obtain their medical history, including the cancer risk. Then, the physician does a guided examination of the places where cancer might be suspected. If a big house had a problem, this would be like asking the homeowners what happened, how they came across the problem, and when they noticed it (Fig. 4.1). Getting as

Fig. 4.1 To quickly identify the faults in a house, it is best to talk to the previous owners. Similarly, checking the patient's medical history is crucial in diagnosing cancer

much information as possible about the issue will help you narrow down your long list of possibilities; it is a process of elimination.

But here is the catch. Not all cancers will alert our body systems immediately. As we saw in the previous chapters, cancers are quite elusive little troublemakers. It often takes a long time for our body's defense system to detect them. For our house, we can imagine a situation where no major appliances are malfunctioning just yet. Cancer patients frequently felt no previous symptoms. It was only a conspicuous blood value or an abnormality in an X-ray or CT scan that led to further clarification and subsequent diagnosis. Some cancers produce various "nonspecific" symptoms that something more familiar might explain. This makes cancers difficult to diagnose because various things could cause one effect or symptom. These symptoms may include sudden weight loss, diminished appetite, fatigue, night sweats, and fever. However, none of these symptoms are very specific nor point directly to cancer since an infection can explain those same symptoms very well. Instead, a constellation of symptoms may be the first hint that a physician will detect.

Furthermore, symptoms that point toward cancer mainly depend on its location in the body. For example, a mass in the colon may not do much except cause occasional constipation. On the other hand, that same mass in the breast may be picked up quickly as a lump that can be felt. The tumor in the colon is deep inside the body, while the tumor in the breast may be closer to the skin's surface. The same is true in the house. A fault deep in the underground wiring may be less evident than a faulty socket where you plug in your phone charger.

Whereas *patient history* is a subjective form of data gathering, an examination is a more objective form. Yet, they are two sides of the same coin. That is why history is so important. A physician uses a patient's history to objectively assess their body to confirm what the patient felt. Then, they could identify signs that may have been missed. For our faulty house, this is like having a look around yourself after talking to the owners (Fig. 4.2). Of course, cancer can also be discovered by chance. For example, if the patient undergoes a chest X-ray for pneumonia, an unexpected 'spot' on a lung might be found.

A cancer *examination* can be general or specific. It can range from your physician checking your whole body for signs of a tumor to focusing on just one part of your body. For the mass in the colon, the physician may check for any palpable masses or abnormalities in your abdomen and in and around your back passage and also obtain the characteristics of your stool. For the mass in the breast, the physician might look for any visible skin changes and asymmetry with the other breast. Then, they will feel the breast to find the lump's exact location, dimensions, consistency, and other characteristics. Like the patient's history, which guides physicians in their decisions, the physical examination steps to take also vary between different cancers. Both the history and the examination help doctors narrow down an impossibly long list of possible differential diagnoses so that a handful of the most likely ones remain. These can then be distinguished from one another by using the tests mentioned later in this chapter.

Comprehensive examinations are always necessary to diagnose cancer. On the one hand, we want to ensure that no benign lesion is taken for cancer; that would be a false-positive result. In this case, we would think the patient had cancer when, in fact, they did not. On the other hand, we would like to ensure that no malignant tumor is taken to be benign; that would be a false-negative result. In that case, we would think the patient did not have cancer when actually they did. Instead, we want to detect a malignant tumor as precisely as possible and then start the best therapy.

Fig. 4.2 Looking for faults in a house. To ensure that we are aware of the damage to the house, sometimes it is not enough to ask, but we need to look around and investigate

General Investigations

Let us return to the analogy of the house. The big house has a problem, and we must find out what it is and where it is. From chatting with the owners, we now know it is a faulty electrical wire. Out of all the issues a house can have, faulty wiring is one of them. Now, we have to find out where it might be, even though there is wiring throughout the house. In a cancer diagnosis, we begin by gathering information on the patient's history and then do an examination. These *investigations* or "tests" determine which diseased or healthy population of cells is likely to have a medical problem that needs to be solved (Fig. 4.3). These tests generally cover either the whole body or one major body region. They help us see *where* the issue is and *what* it might be. In fact, you have probably had at least two of these tests in your life already.

Fig. 4.3 A number of tests are available to identify underlying health issues. These tests can range from simple imaging to genetic testing

4.2   Blood Tests

Blood tests are very common in medical testing. From a simple blood count, called a ***complete blood count (CBC),*** also known as a ***full blood count (FBC),*** we can extract a wealth of information. For our house, which contains a faulty wire, this simple test would probably be like turning the electricity on and off. Like the CBC, this simple check will rarely pinpoint where the faulty wire is located. However, it certainly provides electricians with valuable information they can use to concentrate their efforts further.

In the CBC test, blood is drawn from a person and the sample is sent to a diagnostic laboratory, which counts the number of red cells, white cells, and blood platelets. These numbers can indicate the patient's health status (Fig. 4.4). For example, a massively elevated white blood cell count can reveal the presence of any underlying inflammation and infection, or it can be a sign of a kind of ***lymphoma*** or ***leukemia***, a type of blood cancer. On the other

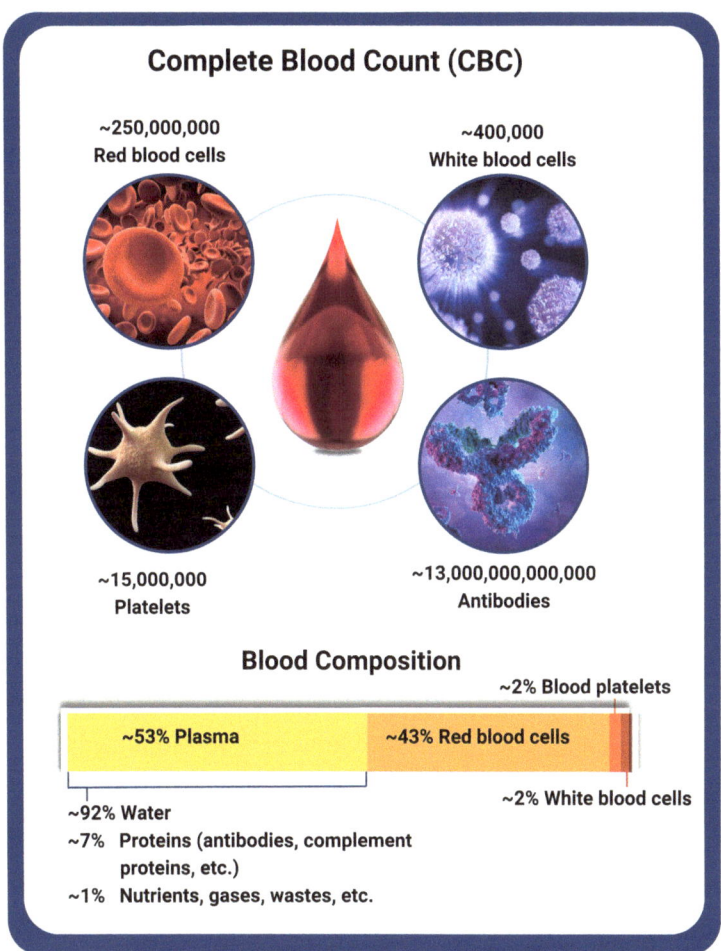

Fig. 4.4 A single drop of blood holds valuable information about our health. Changes in our blood components can reveal underlying conditions that can harm us, such as cancer

hand, a decreased hemoglobin level in the blood (*anemia*) could be a sign of cancer, with symptoms such as pale skin and weakness. However, it could also be a sign of a lack of iron or some vitamin, for example.

CBC is far from the only informative blood test we know of. We have hundreds of tests done on the blood, like the one for *serum electrolytes.* There are also organ-specific tests like the *liver function test, renal function test,* and *thyroid function test.* Many more blood tests have become the go-to initial tests to help physicians quickly get one step closer to the cancerous culprit. These tests can check whether the organs, such as the liver and kidneys, are still functioning well. Knowing this can tell the doctors about the possible contraindications the patient might have against contrasts or dyes used in future imaging tests. Blood test results can take a few minutes to weeks to obtain, depending on the test type. However, as great and cheap as most of these tests are, would it not be more helpful to have something more specific for different cancers?

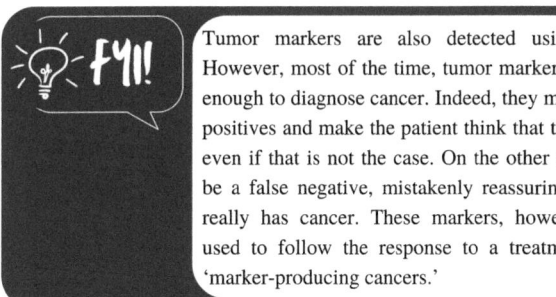

Tumor markers are also detected using blood tests. However, most of the time, tumor markers are not precise enough to diagnose cancer. Indeed, they may result in false positives and make the patient think that they have cancer, even if that is not the case. On the other hand, it can also be a false negative, mistakenly reassuring a patient who really has cancer. These markers, however, are mostly used to follow the response to a treatment, mainly for 'marker-producing cancers.'

4.3   Endoscopy

Increasingly less invasive surgical procedures have revolutionized cancer diagnostics. Endoscopy is a procedure of visualizing the problematic area by using a flexible tube containing a camera and some instruments. If there is a space within the body to examine, there is often an endoscopic way of observing things within that particular organ. The obvious examples are esophagogastroscopy (endoscopy of the food pipe and stomach) and colonoscopy (endoscopy of the colon through the anus), as we saw in the previous chapter. However, such exams are not just limited to the gut and its associated tumors.

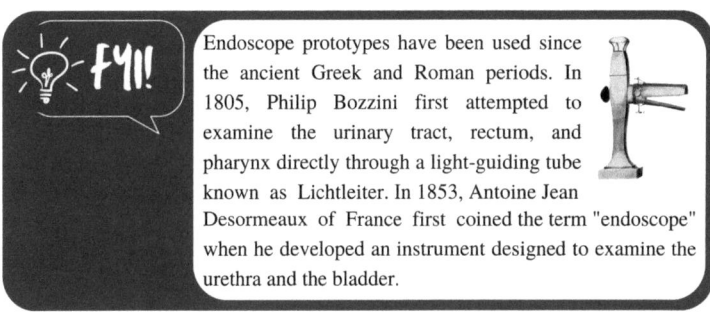

Endoscope prototypes have been used since the ancient Greek and Roman periods. In 1805, Philip Bozzini first attempted to examine the urinary tract, rectum, and pharynx directly through a light-guiding tube known as Lichtleiter. In 1853, Antoine Jean Desormeaux of France first coined the term "endoscope" when he developed an instrument designed to examine the urethra and the bladder.

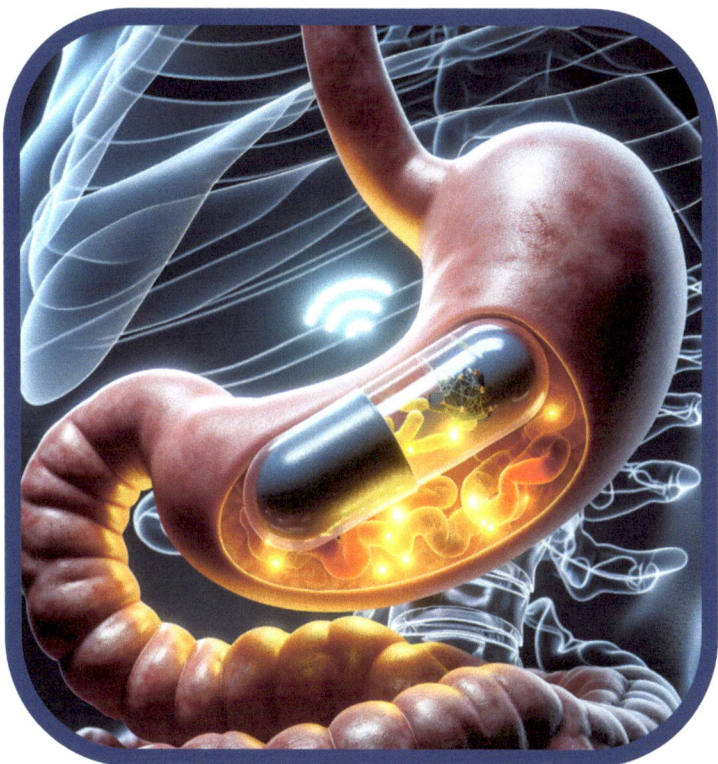

Fig. 4.5 Capsule endoscopy. This procedure involves the use of a tiny wireless camera in the form of a capsule to obtain images of the digestive tract and send them via Bluetooth to a phone

For example, a bronchoscopy is performed to look for tumors in the windpipe. Laparoscopy is done to look inside the abdomen while doing keyhole surgery, and laryngoscopy visualizes the vocal cords. Hysteroscopy is performed to look inside the uterus by way of the vaginal canal. Lastly, capsule endoscopy (Fig. 4.5) can be employed, where a pill with a camera and wireless connection is ingested. The capsule takes pictures throughout your gut before it leaves your body in your stool.

Researchers have developed a new camera capsule in Berlin that can better examine the small intestine. This capsule is designed to show more clearly images of the small intestine. Since 2001, patients have been swallowing pills with cameras that take many pictures of the small intestine as they travel through the body. Until now, these camera pills have had one problem: they take pictures at fixed times, even when the pill is not moving. This means that many images are unnecessary, creating extra work. Therefore, scientists have developed a new capsule that only takes pictures when the pill moves a little in the intestine. This new capsule takes fewer pictures so that doctors can make a diagnosis more quickly. At the same time, in Melbourne, a 2.6 cm long capsule [1] has been developed to measure gas concentrations in the intestine and send the data directly to doctors. The pill has shown that the stomach can produce oxygen and that the large intestine is not entirely anaerobic. The capsule makes it possible to study the activity of intestinal bacteria and could be helpful in the diagnosis of gastrointestinal diseases.

The two projects exemplify the ongoing advancements in endoscopy capsule technology, which are aimed at enhancing the diagnosis of gastrointestinal diseases. However, it's important to note that endoscopy capsules still face significant hurdles before they can be introduced to the market and gain approval despite their technical readiness. We are, therefore, discovering new ways to make tumors visible with endoscopy. The clear advantage of these procedures is that they eliminate the need for a large incision in the area under examination. Endoscopy serves as an excellent alternative to open surgery for visualizing problems, including cancers. These methods are further enhanced when combined with a biopsy, a topic we will delve into later, using the endoscope tube instruments.

Tumors found in all the locations mentioned above can be visualized using endoscopy. The obvious benefit is that these procedures do not require doctors to make a large incision of that particular area to open it up for inspection. Endoscopy is an excellent alternative to open surgery for visualizing problems, including cancers. These methods work even better when combined with biopsy (discussed later) by using the instruments in the endoscopy tube.

4.4   Imaging Techniques

Speaking of taking pictures throughout your gut, imaging techniques have also become as ubiquitous as blood tests. If doctors suspect there is a tumor, they can use imaging tests to determine the ***exact location of the tumor*** and its size (Fig. 4.6). They can also see whether metastases have formed at other sites. Dr. Roentgen's accidental discovery of ***X-rays*** in 1895 in Würzburg revolutionized medicine by allowing doctors to examine the inside of the body without opening it. It is a good example of a noninvasive procedure. Diagnosis of almost everything has benefited from this additional vantage point that medical imaging provides.

Fig. 4.6   The illustration depicts a scenario where an electrician, upon noticing a high electricity bill, meticulously maps out all household appliances to investigate potential issues with wiring or overuse of certain devices. This analogy mirrors the process of doctors utilizing imaging tests to pinpoint the exact location and size of a suspected tumor, as well as identifying any metastases

X-Rays

The most common imaging tool used worldwide is the X-ray machine. It is usually the first-choice scan used for most pathologies, and, quite literally, it is fundamental to the work of bone doctors (orthopedics). Due to their *high frequency and short wavelengths,* X-rays are strong enough to penetrate soft tissue and map out different bony structures in the body. But they are also weak enough to cause no significant damage. Consider them the good old flashlight we use when we are looking for our faulty wire in the house. An X-ray exam may not be the most specific imaging test out there, but it's cheap and gives us immediate results. It is a flashlight that remains tried and true.

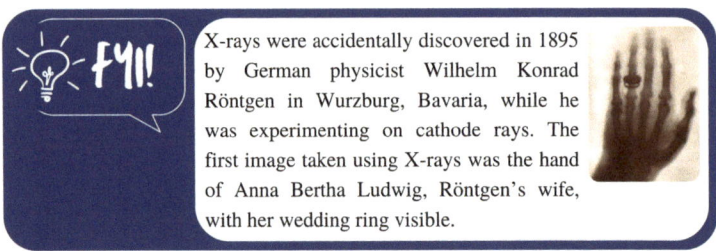

X-rays were accidentally discovered in 1895 by German physicist Wilhelm Konrad Röntgen in Wurzburg, Bavaria, while he was experimenting on cathode rays. The first image taken using X-rays was the hand of Anna Bertha Ludwig, Röntgen's wife, with her wedding ring visible.

Fortunately, bones are not the only place X-rays are helpful. Often, tumors are discovered by chance when we undergo X-ray imaging for other reasons. For example, chest X-rays are equally suitable for locating lung pathologies, including possible tumor masses in the lungs. Abdominal X-rays can also reveal if a bowel-pressuring mass is in the abdomen. That said, pregnant women are advised against having X-rays as the exam exposes the fetus to a high radiation dose. However, in normal circumstances, the risk of radiation exposure is lower than the risk of discovering cancer too late (Fig. 4.7).

During the X-ray procedure, an X-ray beam is directed to a specific body part for a few seconds. An X-ray detector is found on the other side of the patient, where an image will be formed representing the *shadows* of the tissues inside the body. The whole procedure takes less than 15 min to complete (Fig. 4.8). Different tissues have different levels of radiolucency. When X-rays pass through dense areas, they move more slowly and are absorbed more than through thin areas. Bones, for example, reflect the rays, which is why they appear whiter than other tissues against the black background in the X-ray image. Tumors, too, are generally denser than healthy tissue and reflect more radiation. Therefore, a tumor can appear as a bright, conspicuous mass on an X-ray image.

Computerized Tomography (CT)

Today, we have many more options to choose from than traditional X-rays. For example, the 2.0 version of X-rays could be computerized tomography (CT). While traditional X-rays give us a lot of information, they are limited to showing us only hard, radio-opaque structures. They do not provide much of a detailed view of internal organs and other soft-tissue structures. However, CT scans can. CT scans can take an image of an area of the body, layer by layer, and easily identify the exact location of a tumor, its size, and even its blood supply. A CT scan shows us tissues and organs without the overlay of other organs that would otherwise hinder our view.

Fig. 4.7 Patients often fear undergoing scans, even when medically necessary, due to concerns about radiation. However, they may not realize that they willingly expose themselves to radiation when flying on vacation, too. The purpose of the text and accompanying image is to highlight this paradox and illustrate that flying also carries risks. For instance, comparing the radiation exposure from a typical pelvic CT scan (which ranges from 2 to 10 millisieverts) to that experienced during a transpacific flight between Los Angeles and Japan. Passengers on such flights are estimated to receive significantly lower radiation exposure, typically ranging from 0.01 to 0.15 millisieverts, compared to the higher doses associated with a pelvic CT scan. However, these doses multiply with many flights

During the CT scan, the X-ray beam rotates around the body, revealing far more detail. A CT scan is a computer-generated image from the information gathered by hitting the body with X-rays that are stronger than those in conventional radiographs. The rays also hit the body at specified angles to create sectional images called *slices*. The X-ray data is then sent to a computer that interprets it. The computer then processes the images from these slices. The

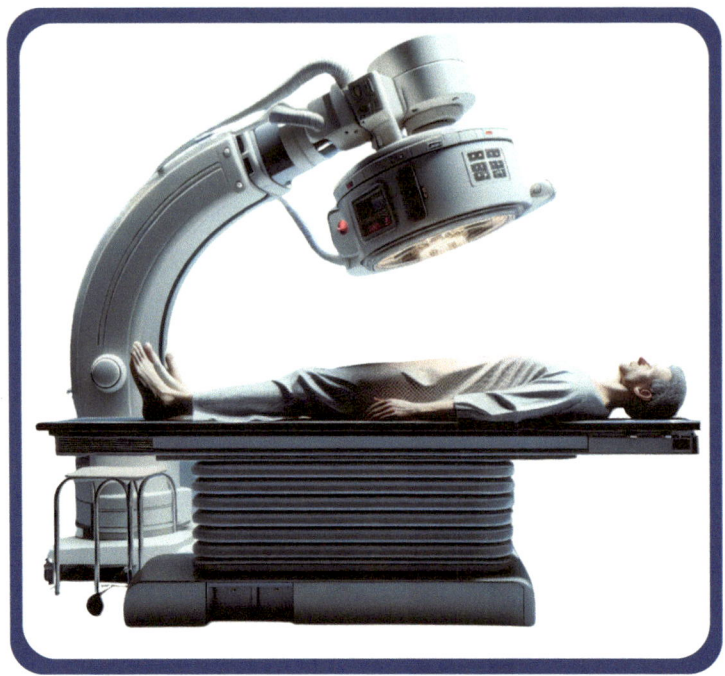

Fig. 4.8 X-ray Imaging. During this test, a beam of X-rays targets a body part. An X-ray detector is found on the other side of the patient, which creates images of the tissues the X-rays pass through

output is a collage of a two-dimensional *image* on a monitor showing the inside of the body in much more detail than standard X-rays [2] (Fig. 4.9).

CT scans can also be enhanced. We can do that by adding an iodine-based radio-opaque dye to the patient's blood. This increases the contrast between the different types of tissues. Patients either drink the dye or have it injected into their veins. This enhancement is routinely used to check the caliber and efficiency of blood vessels in different organs, like the liver and the heart. The contrast material highlights areas that are affected in the body and can help doctors distinguish blood vessels from other structures. This painless procedure can take about 10–30 min to complete, depending on the area that is imaged.

As with X-rays, however, pregnant women are advised to opt out of this imaging test since CT scans rely on higher-frequency X-rays for their remarkable resolution. Therefore, the downside to this test is that it exposes people to more radiation than an X-ray machine would. In addition, the dye used as

Fig. 4.9 CT scan. The patient remains motionless on a table during the treatment while it slowly moves through a doughnut-shaped scanner that emits rotating X-rays all over the patient's body

the contrast material used in CT scans can harm people with impaired kidney function. Therefore, it is essential to have everything checked beforehand. Nevertheless, the benefits of CT sometimes outweigh the risks, and a non-radiation alternative is even available in certain cases.

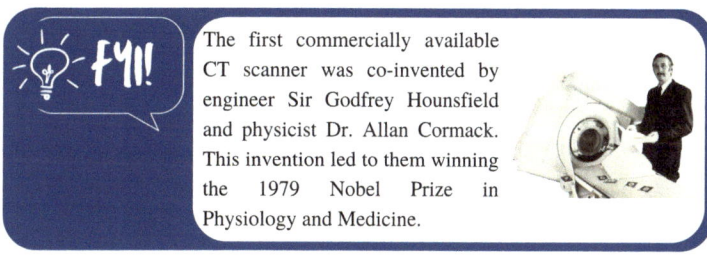

The first commercially available CT scanner was co-invented by engineer Sir Godfrey Hounsfield and physicist Dr. Allan Cormack. This invention led to them winning the 1979 Nobel Prize in Physiology and Medicine.

Ultrasound (Sonography)

In medicine, various imaging techniques help diagnose and treat diseases. Unlike X-rays and CT scans, *sonography* uses ***ultrasound waves*** instead of radiation to obtain images. During an ultrasound, a transducer sends ultrasound waves to a tissue (Fig. 4.10). Different tissue types reflect these waves differently, and these reflections are converted into an image. The sonographic image, or sonogram, shows the outlines of the organs and the tissues.

Sonography is an important cancer diagnostic tool. For example, it can detect a lump in the breast, easily distinguishing a solid tumor from a fluid-filled cavity or cyst. Different tissues reflect the sound differently; it depends on whether the tissue is solid or contains liquid. The image we obtain is influenced by the tissue type, as each reflects sounds differently. Unfortunately, this

Fig. 4.10 Ultrasound imaging. (A) In the process of ultrasound imaging, a sonographer uses a device to send sound waves into the body, capturing the returning waves and sending the data to a computer to create images. Submarines operate similarly, using ultrasound pulses to determine the distance and location of target objects. This method provides valuable information for their operations

technique cannot be used for organs filled with air, like the lungs and intestines. Therefore, in some cases, ***endosonography*** is the diagnostic tool to use. Here, the transducer is brought inside the body through an opening in the body close to the suspicious tissue. For example, it may be put through the mouth to examine the esophagus. Thus, the sound is sent from the inside, not through the skin. This procedure is used in gastroscopy and colonoscopy.

Some of the advantages of ultrasound have included its easy access, as it is available almost everywhere. It is also noninvasive and can be done quickly in less than 20 min. The doctor can thus immediately evaluate the images on the monitor. Most importantly, it does not expose patients to radiation. However, if a more detailed image is required, then another widely available option is available.

Combining artificial intelligence (AI) with ultrasound technology has opened up new possibilities in the field of medicine [3]. Recent advancements in artificial neural networks and deep learning have greatly improved ultrasound technology. AI can enhance the accuracy of ultrasound examinations, particularly in breast sonography, potentially replacing or complementing procedures such as X-rays. This could result in cost savings and eliminate the need for invasive methods or those involving high radiation exposure. In breast cancer diagnostics, automated scanners capture 3D ultrasound datasets, which are then analyzed by AI to detect any signs of malignancy. The results are comparable to those obtained by experienced specialists. AI-supported ultrasound can also enhance mammography screening by more accurately detecting tumors in dense glandular tissue, potentially identifying around three additional carcinomas in 1000 examinations. Despite its advantages, there is a training challenge associated with ultrasound due to the high level of expertise required and the steep learning curve. AI-supported automated screenings could assist by sorting inconspicuous images, reducing the workload. Furthermore, AI can also serve as a training tool by supporting young doctors in analyzing ultrasound images.

Sonography not only aids in diagnosis but is also utilized in the treatment of patients. In the United States, a groundbreaking procedure known as histotripsy has been approved for the treatment of liver cancer [4]. This innovative technique uses ultrasound waves to eliminate tumors in a minimally invasive manner without any side effects, as it bursts microbubbles to destroy tumor cells, which are then eliminated by the immune system. Healthy tissue remains unharmed as the ultrasound precisely targets the tumor. Histotripsy offers patients an alternative to radiation and chemotherapy with fewer medication issues and a shorter recovery time than surgery. The treatment is not only effective but also less burdensome.

Magnetic Resonance Imaging (MRI)

The MRI scanner has become a beacon for doctors looking for a radiation-free alternative to CT scanners. MRI scans are well-known for their high-quality images. They can distinctly differentiate healthy from abnormal tissues. They can also produce cross-sectional images like a CT but from different angles. Soft tissues, such as muscles, fat, and blood vessels, can be visualized and differentiated better with MRI than with CT. This enables spinal cord or brain tumors to be detected.

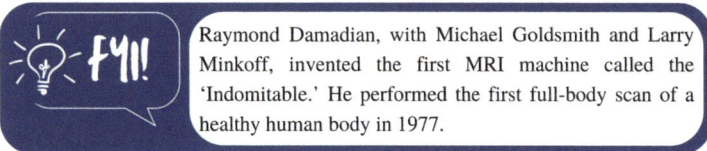

Raymond Damadian, with Michael Goldsmith and Larry Minkoff, invented the first MRI machine called the 'Indomitable.' He performed the first full-body scan of a healthy human body in 1977.

MRI works on the principle of magnetic resonance imaging, as ***superconductive magnets and radio waves*** produce images (Fig. 4.12). Soft tissue resolution is better in MRI images than in CT images. However, MRI scanners are very expensive, and the time it takes to obtain a full set of MRI images can seem long (30–60 min). In MRI exams, a dye can also be used as contrast media, administered in advance. However, unlike enhanced CT exams, the

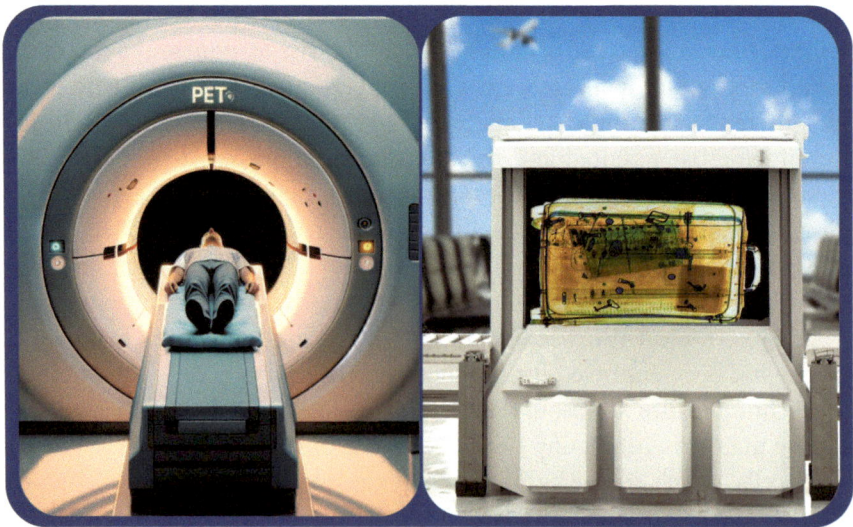

Fig. 4.11 PET scan. As in a CT scanner or MRI scanner, patients must lie very still during a PET scan as the narrow-padded table they lay on slides slowly back and forth through the machine's hole

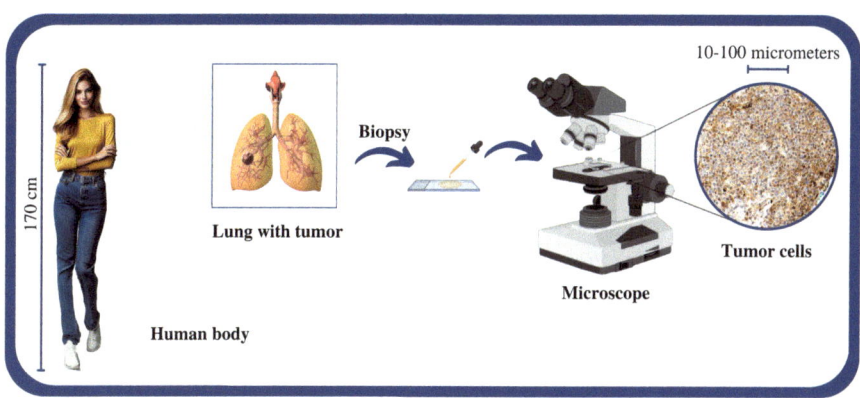

Fig. 4.12 Histology. Human tissue samples are extracted from the tumor through biopsy, stained and studied through a microscope

dye in MRI exams does not damage the kidneys. That said, dialysis patients should be cautioned from taking MRI contrast media that contain gadolinium.

Radiological exams are complimentary. Some things, such as the lungs and bone calcifications, are better seen with a CT scan than with an MRI scan. Furthermore, because it uses strong magnets to scan the body, MRI cannot be used for people with metallic implants like pacemakers, hip replacements, insulin pumps, and nerve stimulation devices. For these patients, a CT scan is the only option. Although MRI does not expose patients to radiation, its effects on the embryo have not yet been thoroughly researched. It is thus not advisable for a pregnant woman to have an MRI scan during the first 12 weeks of pregnancy. Patients' ears also have to be protected during the procedure since the device makes loud noises. Finally, claustrophobic patients might also need help overcoming their fear as they lie in a closed tube and are not allowed to move during the test.

However, despite these concerns, MRI scans are still useful in detecting cancer. They can detect soft tissue masses much better than CT scans. If CT scans are the flashlights we use to look for that faulty wire, then MRI scans are the LED versions of those flashlights.

In Dresden, a new device was presented that combines whole-body MRI with proton therapy. This is a decisive step in cancer treatment, as it enables more precise tumor irradiation and makes treatment gentler. The technology detects changes in the tumor and adjusts the volume of radiation accordingly. The development was challenging as both systems use magnetic fields that potentially interfere with each other. However, researchers found a solution to effectively combine both technologies. MRI imaging opens up new avenues for proton therapy. Collaboration between different disciplines and a focus on innovation is crucial for cancer therapy. The launch of the prototype shows that research and medical care are setting new standards [5].

Positron Emission Tomography (PET) Scan

PET scans produce images that are lower in resolution than those from CT or MRI, but they do it in style (Fig. 4.11). CT scans offer clear and detailed pictures of the body's internal organs, while PET scans work efficiently as cancer highlighters. They can be considered the equivalent of black light, where the insulation of the faulty wire (provided it is made of the right material) would glow under the metaphorical black light of the PET scan. A PET-CT is particularly useful for finding an original tumor and any metastases that may be present. It is also used to check the success of therapeutic measures such as chemotherapy. It can be used in follow-up care to confirm a suspected metastasis or to plan a new operation [6, 7].

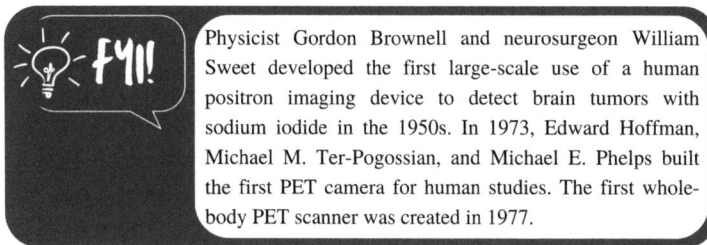

Physicist Gordon Brownell and neurosurgeon William Sweet developed the first large-scale use of a human positron imaging device to detect brain tumors with sodium iodide in the 1950s. In 1973, Edward Hoffman, Michael M. Ter-Pogossian, and Michael E. Phelps built the first PET camera for human studies. The first whole-body PET scanner was created in 1977.

Nuclear Scan (Scintigraphy)

Nuclear scans tag cancers at close range. The body is not irradiated at a high level but rather briefly becomes a source of radiation and, thus, an image source. Nuclear scans are particularly useful in detecting functional cancers. Functional cancers like thyroid cancers occur when cancer produces an active compound like an enzyme or when a hormone using raw materials is taken up. The raw material in this test is weakly radioactive, unlike the tracer for PET scans, which is more radioactive [8]. The tracer used depends on the cancer under suspicion.

The nuclear procedure takes advantage of the fact that some substances preferentially accumulate in specific tissues or organs. Take thyroid cancer, for example. In this case, iodine will be the tracer. Weakly radioactive iodine will be injected into the blood for a nuclear scan looking for suspected thyroid cancers. The iodine will be absorbed primarily by the thyroid gland. An image of the thyroid is then taken a little while later by using a special *gamma camera* to see how much of the radioactive iodine was taken up by the gland to produce the thyroid hormone (Fig. 4.14). Through the gamma camera, we can see precisely where the radiation is emitted in the thyroid gland. The image will show us metabolic activities that are higher in tumor cells than in healthy cells. Elevated radioactive iodine uptake, combined with the history, examinations, and tests indicative of a thyroid tumor, would get us amazingly close to a final diagnosis here.

All in all, this kind of scan takes about 30–60 min. The procedure can also identify bone tumors and metastases. There, the principle is the same, but radioactively labeled substances are administered instead of iodine. These substances, called poly-phosphonates, are preferentially deposited in the bones. As malignant cells have a higher metabolic activity than benign ones, the concentration of these radiolabeled substances in bone tumors is higher than in healthy bone tissue. This is how we can find bone cancer, for instance. We can think of nuclear scans as spying. They send a *spy* into a gang to find out how it works and who its members are. Nuclear scans are currently the most effective technology we have for obtaining noninvasive images, which identify the location, extent, and often even the type of cancer under suspicion. This speeds up our process of elimination. However, the gold standard in crime solving is not to take pictures of crime scenes but to catch the culprits in the act. Therefore, the best evidence we can hope for when diagnosing cancer is to catch cancer red-handed. This is what nuclear scans help doctors do.

Very recent anticancer treatments utilize this technique to administer tracers designed to irradiate the cells in their vicinity (the goal is to irradiate them, not just "mark them for examination"). These treatment tracers are much more radioactive than the tracers used to identify cancerous lesions. By injecting radioactive lutetium linked to the tracer PSMA, which is a molecule particularly expressed on the surface of prostate cancer cells, we can target metastatic prostate cancer cells that are known to attract this tracer.

While these tests are important, they do not provide a definitive diagnosis. Only a biopsy can absolutely confirm the presence of cancer, specify its type, and sometimes determine its aggressiveness, which we will discuss in the next paragraph.

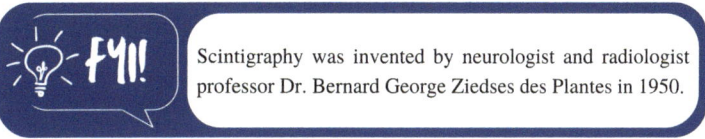

FYI! Scintigraphy was invented by neurologist and radiologist professor Dr. Bernard George Ziedses des Plantes in 1950.

4.5   Histological Examination

As discussed in previous chapters, cancer cells have many ways of maneuvering around the loopholes of our immune system. They can thus plant themselves firmly in a position to grow, divide, and invade other neighboring cells and tissues. We know this because we have caught cancer cells in the act, both in tumors and by microscope. Cancer may be diagnosed using the techniques discussed earlier, either alone or in combination. However, definite proof comes from a biopsy when we extract cancerous cells from the body (Fig. 4.12).

4.6   Biopsy

In medicine, a biopsy is like catching cancer red-handed at the crime scene. When we have a suspicion about *blood* or a *tumor mass,* we can use various techniques to examine them under microscopes on specially prepared slides. For our faulty wiring problem, this is like using electrical testers and voltage meters to see if a wire is hot. If it is, we may see it glow red-hot, send sparks, and short a fuse before our very eyes. Biopsies are the gold standard test for diagnosing any condition involving cells directly, including most cancers known to physicians.

A pathologist uses a microscope to analyze cell features, such as cell numbers, cytoplasmic rations, membrane invasions, and cell colors, to figure out what kind of cancer they are. But the trick here is to get a sample that is suitable enough for analysis. Unfortunately, we can miss the target when doing a biopsy, which is why we do not routinely use biopsies alone for diagnostic tests (Fig. 4.13). After we have obtained a thorough patient history and done a physical examination, we need the right combination of blood and imaging tests to give us the crucial piece of information we need for biopsies: where is the mass that needs to be tested?

Once we know accurately where to look for the suspected cancer, it is just a matter of getting a sample large enough to look at under a microscope. However, depending on where the cancer is, different methods need to be used to get a viable sample. In most cases, biopsy procedures take around 1 h, and the results can be available within a few days or 2 weeks. In fact, cell coloration, preparation for genetic tests, and final diagnosis can take between 2 and 4 weeks. Cancer researchers are thus now using artificial images to try to

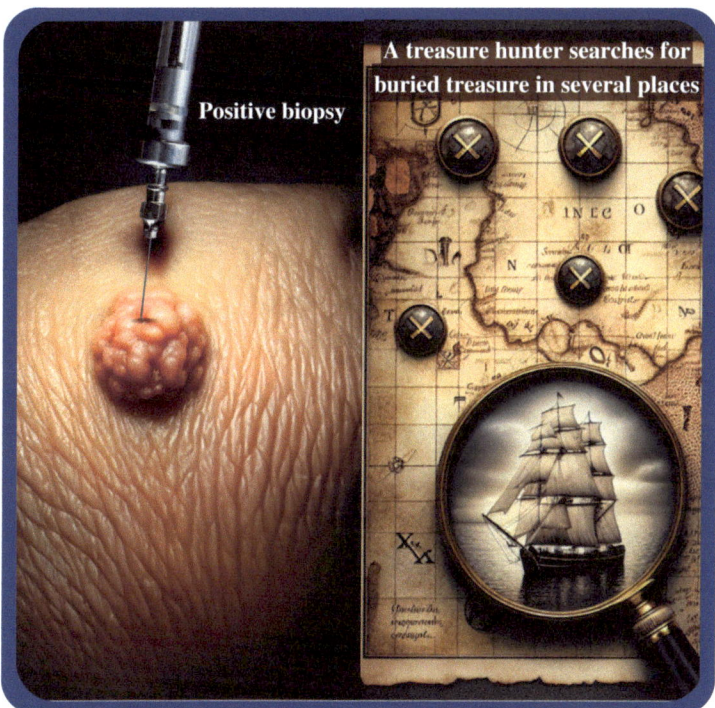

Fig. 4.13 Missing the target. Sometimes, a repeat biopsy is needed in a benign finding to rule out the possibility of a missed puncture. A repeat biopsy can be likened to a treasure hunter who needs to dig in several places before eventually finding a buried treasure. While cancer is not a treasure, avoiding false-negative findings is very important

accelerate this process. They digitize pathology slides, convert them into images, and apply algorithms to them so that cancer cells can be detected automatically one day. If this comes to fruition, cancer will be diagnosed more quickly [9]. But at the moment, these are dreams for the future.

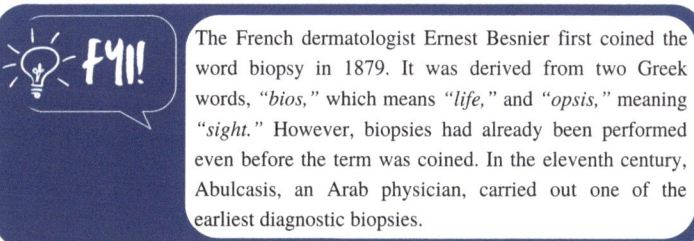

The French dermatologist Ernest Besnier first coined the word biopsy in 1879. It was derived from two Greek words, *"bios,"* which means *"life,"* and *"opsis,"* meaning *"sight."* However, biopsies had already been performed even before the term was coined. In the eleventh century, Abulcasis, an Arab physician, carried out one of the earliest diagnostic biopsies.

Needle Biopsy

For masses close to the skin, there is a good chance of hitting the target in the first go. Indeed, samples of the mass can be obtained simply by feeling the lump, sticking a *syringe* needle into it, and retrieving some cells (Fig. 4.14). Needle biopsies are quick and have less scarring than other options. For thyroid masses, a breast lump, a nodule on the skin, or a swollen lymph node, needle biopsies can get the job done for most of these *superficial solid tumors*. However, sometimes, the sample size may not be big enough to give us an adequate picture to work with. To get bigger samples, we may have to do surgical biopsies.

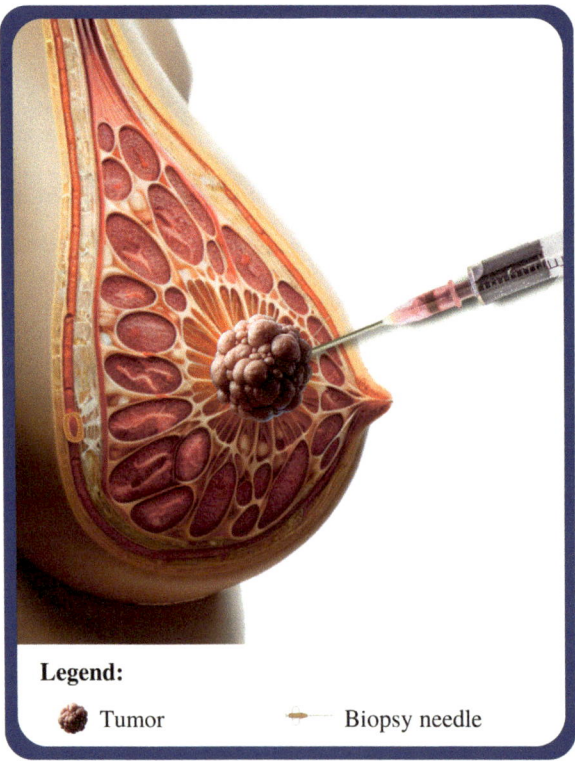

Legend:

🔴 Tumor   ➤ Biopsy needle

Fig. 4.14 Needle biopsy. During this procedure, an anesthetic may numb the area to be biopsied. A doctor will guide a needle through the skin to the tissue of interest and collect a sample of cells when the needle is withdrawn. This process may be repeated until enough cells are collected. Patients may experience pressure and discomfort in the area involved

Surgical Biopsy

Surgical biopsies are different techniques used surgically to obtain a sample of a suspected tumor mass. Terms like *incision* and *excision* biopsy refer to different ways of surgically removing a piece of the mass that is large enough so that enough slides can be made for a histological study. Surgical biopsies help us retrieve a larger sample of the mass than needle biopsies and enable doctors to dig deeper into the body for a sample than a syringe needle could. But the surgery itself is never risk-free.

Blood Smear and Cytometry

Not all cancers are solid. *Blood cancers* can be suspected from blood tests alone. However, a biopsy is still required to prove that the massive white cell count comes not from infection but from leukemia. Blood smear and cytometry mean that a blood sample is taken and then prepared with special stains to be studied under a microscope. The nature and appearance of the cells in question can help medical staff classify and differentiate different blood cancers. Usually, running the test only requires a sample of about 20 ml of blood and 2 ml of organic solvent. For example, markedly elevated white blood cell counts may represent a type of leukemia called chronic myeloid leukemia (CML). The appearance and maturity of the white blood cells studied histologically determine which one of the six varieties of CML we have. This process can be compared to using a soil probe to obtain a soil sample and scan its components. The result can be used to determine whether the soil is fit for its intended purpose, such as constructing a building on the site.

Liquid Biopsy

Liquid biopsy is an innovative procedure with the potential to revolutionize cancer diagnosis and treatment. Unlike traditional biopsies that require tissue sample removal, liquid biopsy looks for solid cells or DNA in the blood. In other words, we're examining the tumor DNA circulating in the blood. This technique is increasingly used to identify circulating tumor cells (CTCs), cell-free DNA (cfDNA), and circulating tumor DNA (ctDNA) in blood samples, which could lead to significant advancements in cancer research (Fig. 4.15).

The analysis of ctDNA is especially promising as it can provide insights into the progression of tumors. For example, an increase in ctDNA levels may

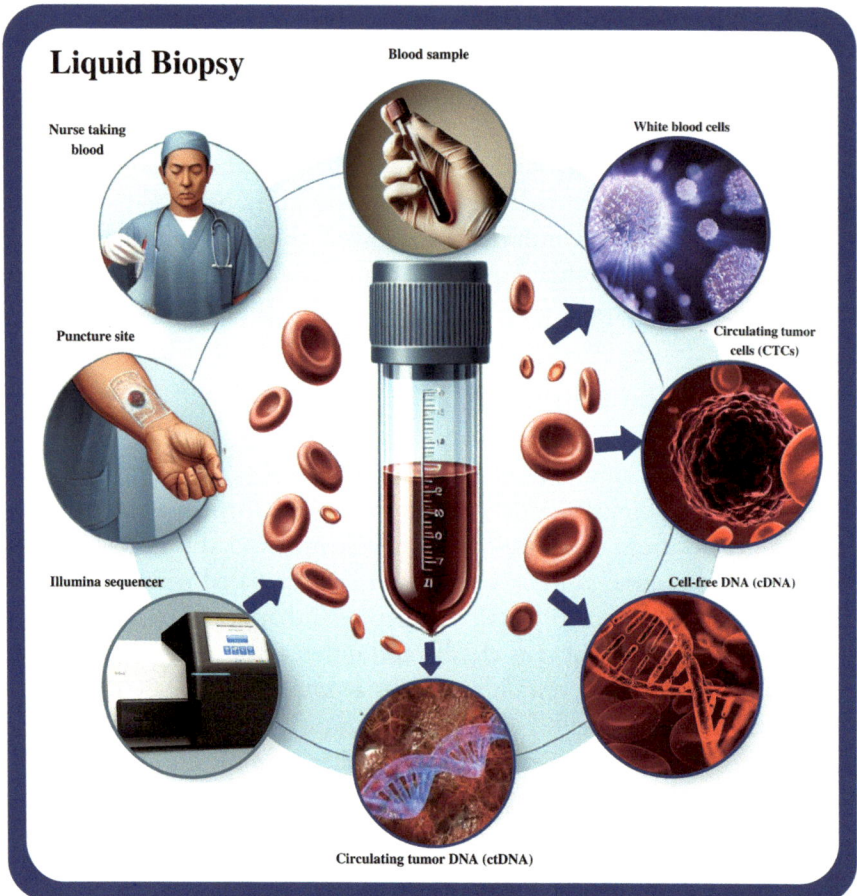

Fig. 4.15 Liquid biopsy. Different components from a blood sample are measured to indicate health status and diagnose possible cancers

indicate worsening cancer. Although this method is not yet commonly used in daily practice, it is actively researched and could soon play an essential role in cancer care. Two recent studies provide an understanding of liquid biopsy.

One study in Kingston demonstrated that the early disappearance of ctDNA from the blood in advanced lung cancer patients is associated with prolonged survival when treated with the checkpoint inhibitor pembrolizumab. The study aims to determine the prognostic significance of the disappearance of ctDNA and the optimal timing for the test. The results of the liquid biopsy correlated well with X-ray controls and were able to detect molecular remissions after just 2 months, significantly earlier than X-ray images.

Another study in Shanghai indicated that analyzing ctDNA methylation in the blood of colorectal cancer patients could detect relapses early, improving postoperative treatment. Typically, follow-up care involves measuring CEA in the blood or performing a colonoscopy. Researchers discovered that ctDNA in the blood indicates disease progression. In a pilot study, they were able to detect early relapses using ctDNA methylation markers.

A more recent study investigated the relationship between ctDNA methylation and colorectal cancer recurrence. The combination of ctDNA status and CEA test improved the risk assessment. After chemotherapy, patients with ctDNA had a shorter relapse. When a patient with advanced colorectal cancer is found to have an oncogenic mutation of the EGF receptor, this leads to continuous activation and aberrant proliferation of cells, which forms the foundation of the disease. Cetuximab, an antibody, can effectively attach to this receptor, inhibiting its excessive activity and decelerating the progression of the tumor [10]. Cetuximab selectively targets the patient's unique genetic mutation.

It's important to remember the breakthroughs in colorectal cancer treatment, particularly the potential of analyzing ctDNA methylation in the blood of patients to detect relapses early and enhance postoperative treatment. This finding emphasizes the vital role of medical professionals, researchers, and oncologists in advancing our understanding of the disease. The combination of ctDNA status and CEA test was shown to significantly improve risk assessment, providing valuable insights for patient care.

4.7   App Diagnostic

The use of digital technologies, algorithms, and artificial intelligence (AI) in modern diagnostics is revolutionizing the detection and treatment of diseases, particularly cancer. For instance, researchers in Sydney have created AI-powered smartphone apps that can diagnose skin cancer with a level of accuracy on par with dermatologists (Fig. 4.16). While doctors excel in devising treatment plans, this breakthrough underscores the potential of AI in medical diagnostics. In a study, two AI algorithms were tested in skin cancer centers and compared with the diagnoses of doctors with varying levels of experience. One algorithm performed as well as highly experienced doctors and surpassed less experienced ones. Although doctors remain superior in treating skin cancer, the potential of AI in medical diagnostics is evident. The scientists emphasize the value of AI apps in diagnosis while highlighting the need to integrate AI into treatment planning and to test AI applications in

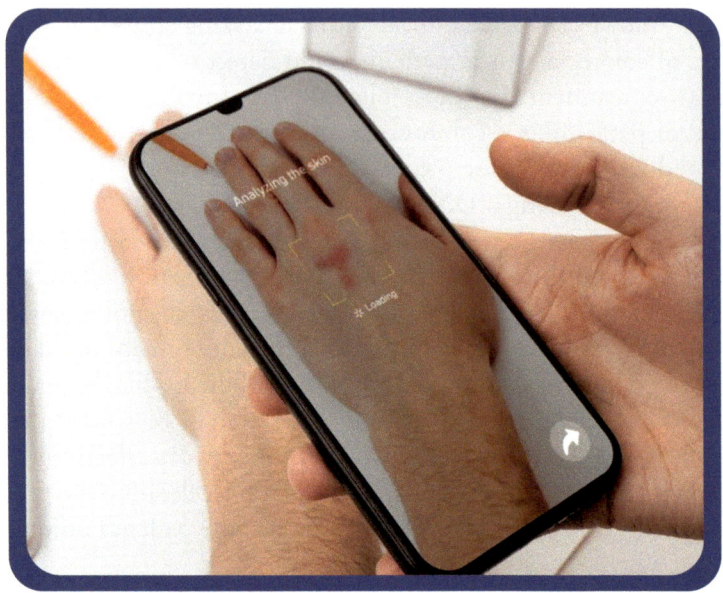

Fig. 4.16 The image shows a person using a smartphone to analyze a skin condition on their hand. The phone screen shows an app user interface focusing on a red area, possibly a rash or irritation, while a message indicates that the skin is being analyzed. In the future, such apps will also be able to diagnose skin cancer

practice rigorously. These advancements in AI-assisted diagnostics offer great promise for enhancing early cancer detection and treatment, ultimately saving lives.

4.8   Genetic Testing

We began our journey into diagnosing cancer from the outside by looking inside and moving closer, step by step, to the primary culprit. With each step in data gathering, we have delved deeper, moving from the general history and symptoms to watching cancer cells go about their day under the microscope. The focus has thus become narrower. However, the journey does not end there. If we dive deeper and go beyond the walls of individual cancer cells, we find ourselves looking face-to-face at DNA, the mastermind. DNA contains the building blocks of all living cells, even cancer cells. DNA is the source code for how life operates and why our cells behave the way they behave. Cancer cells live according to their DNA programming. For many of us, our DNA answers the question of whether we are at risk of getting cancer.

Often, it is helpful to go deeper due to cancer's diversity. Going deeper can help identify a specific cancer type, especially when there is a lack of clinical signs. By going deeper, we can identify genomic problems and mutations such as translocations and inversions. Any mutation we identify could determine whether there is a risk of cancer recurring or a risk of developing another cancer type. In addition, identifying the genetic changes can guide cancer treatment [11]. Using genetic testing for mutated DNA segments is the pinnacle of a cancer diagnosis. Find the malfunctioning code, and you can identify what exactly happened and where to go from there. If we learn what our DNA reveals and what that revelation means for cancer cells, then we can answer the next logical question: why is this cancer?

Karyotyping

Some genetic abnormalities are not localized to one single gene like the *BRCA*. In fact, an aberration may occur on an entire chromosome. For example, Down syndrome is a genetic condition where a person has three copies of chromosome 21 instead of two. Karyotyping can help find such aberrations. It continues to be the gold standard in cancer cytogenetics, which **tests a cancer cell's entire genome** to determine *why* it acts *how* it acts. Karyotyping chromosomes in cancer cells, starting with the original mass, helps us determine genome-wide defects. It also predicts how the patient may respond to different treatments. This method is commonly used for myeloma, some lymphomas, and leukemia. Karyotyping is like playing a Mikado game. To free the sticks from the jumble, each stick must be taken up in a particular order without moving or touching others. Similarly, to detect the right karyotype in genetics, the lab technician must put all the chromosomes in order, from chromosome 1 to chromosome 23. Then, the lab technicians can look for anomalies like trisomies or translocations (Fig. 4.17).

Fluorescence In Situ Hybridization (FISH)

While PET scans highlight cancer, FISH scans highlight mutated genes. In the cytogenetics world, FISH works by making **mutated genes easier to recognize.** FISH does this by enhancing color. FISH scans target specific genes whose sequences are already known. In this technique, from the 1980s, blood or cancer tissue samples are extracted. Afterward, *fluorescent dyes* are attached to DNA probes to target specific chromosomal locations within the nucleus.

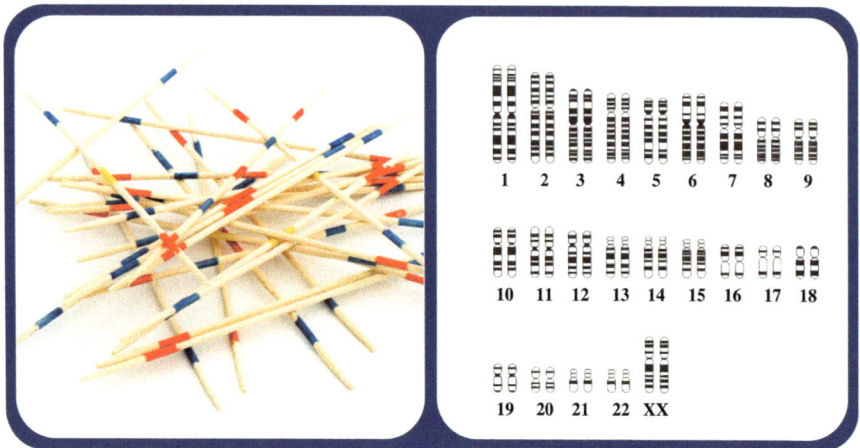

Fig. 4.17 Mikado and karyotypes. Both of these activities require each stick or chromosome to be taken in a particular order to unjumble and identify any anomalies

This produces colored signals that fluorescent microscopes detect, which confirms a mutant gene is present [12]. This scenario can be compared to fishing, where one needs a particular lure to attract a specific fish species.

The advantage of FISH is that it directly identifies mutant genes and chromosomes. However, FISH is limited by the fact that the number of known mutant genes is finite. Then, there is the short life of the fluorescent signal itself, which also limits FISH. It is why FISH can be complementary to other tests rather than used alone. The good news is that FISH can diagnose breast cancer and certain types of lymphoma [13, 14]. For example, chronic myeloid lymphoma results from the translocation or "swapping" of some parts of chromosomes 22 and 9. When the parts change places, the *BCR-ABL1* oncogene is formed. This translocation is also known as the famous *Philadelphia chromosome,* discovered by David Hungerford in 1959 [15].

Genomic Sequencing

The Human Genome Project began in 1990. Sequencing the entire genome of a cell was a massive undertaking then. However, when the project finished in 2003, we had around 90% of our DNA in decoded form. By 2006, the entire human genome had been sequenced. All the secrets to cancer cells, their causes, their behavior, and their ultimate solutions all lie within the double helix structure of our DNA. Detecting cancer using DNA sequences will improve in the future as DNA sequencing becomes quicker and more

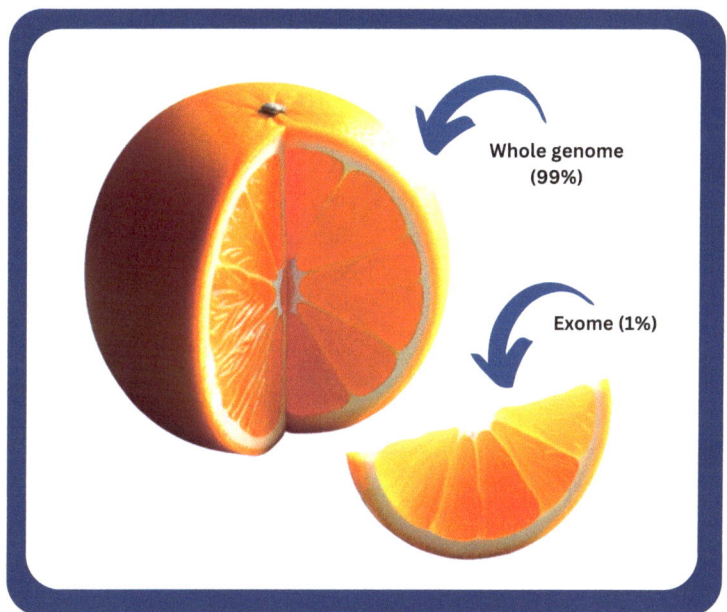

Fig. 4.18 Exome in a genome. Exomes are important sections of the genome. However, they comprise only 1% of the whole genome

affordable, whether as a state effort or by private companies like 23andMe, detecting cancer by using DNA sequences will improve in the future.

Different methods can be used to sequence genomes. ***Tumor genome sequencing*** can sequence the entire genetic material. ***Exome sequencing*** is done only where the sections translated into proteins are read (Fig. 4.18). ***Gene panel sequencing*** is done when only the sections of the genetic material involved in cancer are sequenced. Otherwise, a ***single genetic test*** can be done [16].

Precision Techniques

Precision oncology is a groundbreaking trend in cancer diagnostics that offers unparalleled accuracy in identifying cancer DNA and mutations, revolutionizing the field. This advancement is pivotal as it paves the way for highly effective treatment methods. Techniques such as the creation of a personalized genetic profile for each patient are instrumental in this process. Genetic profiling of a patient helps doctors to develop personalized treatment plans, fundamentally changing the way we look at cancer and how we treat it. Drugs can now be designed to target only cancer cells without harming healthy cells.

If we know the genetic characteristics of cancer cells, we can tailor treatment to the individual. New techniques are being researched to speed up and improve each cancer patient's genome profiling. Sequencing the human genome, which once took 16 years, can now be done in a fraction of that time, and next-generation sequencing technology enables the rapid identification of all relevant tumor DNA variants, including complex biomarkers. Precision oncology is an adjunct to established therapies in clinical trials and off-label therapies. The results are discussed in molecular tumor boards (MTBs), which interface between diagnostics and clinical practice. Despite the advantages of precision oncology, only around a third of patients receive the recommended therapy to date due to a lack of clinical trials, and health insurance companies often need to cover the costs of unapproved therapies. The refusal to cover expenses is based on different views of molecular evidence that do not meet the requirements of evidence-based medicine. Precision oncology is a promising approach that could revolutionize cancer treatment by tailoring treatment to the genetic characteristics of the individual patient. By combining next-generation sequencing and personalized therapies, doctors and scientists can work towards increasing the effectiveness of cancer treatment and improving patients' quality of life. However, there are still many challenges to overcome before precision oncology becomes accessible and affordable for all patients.

Sanger Sequencing

Sanger sequencing is a first-generation method to ***detect mutations within a defined area of the DNA.*** To detect cancer, Sanger Sequencing makes it possible to identify the culprit gene. That information is useful in developing drugs that target mutated genes. Although many techniques have been superseded by parallel-sequencing techniques like next-generation sequencing (NGS), Sanger sequencing is still used as a benchmark that newer gene sequencing techniques are compared to.

Next-Generation Sequencing

Next-generation sequencing (NGS) has revolutionized genetic analysis by significantly enhancing the speed and scalability of DNA and RNA sequencing. This technology allows researchers and laboratories worldwide to analyze multiple gene alterations in a single comprehensive analysis, making it indispensable in genetic research.

In the realm of cancer diagnosis, NGS plays a transformative role by enabling rapid and thorough sequencing of the genome, including specific regions implicated in cancer. For instance, in melanoma, NGS can detect mutations such as BRAF V600E in melanocytes. Its ability to simultaneously sequence and quantify multiple DNA sequences facilitates the identification of crucial genes for precision therapy, which is crucial given the diverse nature of tumors.

A prime example of NGS's potential lies in whole genome sequencing (WGS) [17, 18], which promises to revolutionize cancer treatment. WGS examines an individual's entire genome, providing profound insights into the molecular mechanisms of cancer. Initiatives like the 100,000 Genomes Cancer Program leverage WGS to uncover genetic variants linked to various cancers, integrating these findings with clinical data to optimize treatment strategies and enhance survival rates.

While integrating WGS into routine clinical practice presents logistical, technical, and economic challenges, the comprehensive genetic landscape provided by WGS is invaluable for informed therapy decisions. NGS and WGS have vast potential in cancer treatment, promising to improve therapy efficacy and patient outcomes significantly.

In summary, NGS enhances genetic sequencing capabilities with unprecedented speed and scalability. Other sequencing methods are like the iPhone 4 compared to the NGS's iPhone 14, as the early iPhones had fewer features than more recent iPhones (Fig. 4.19). Its applications in cancer diagnosis, including detailed genome and cancer DNA sequencing, highlight its transformative impact on precision medicine. Despite challenges, the integration of NGS and WGS holds immense promise for advancing cancer treatment through personalized therapeutic approaches tailored to individual genetic profiles.

Researchers have developed a new AI model that can predict genetic biomarkers [19] (*MSI, BRAF, KRAS*) in colorectal cancer. The model is based on a transformer neural network and shows high accuracy, comparable to clinical tests on biopsies. The approach uses deep learning to analyze tissue samples, which is particularly advantageous for countries with limited resources. The model was trained with over 13,000 patient samples from different countries. It achieved a sensitivity of 99% in predicting MSI on surgical resection specimens and also showed high accuracy on biopsy tissue. This technology can speed up the diagnostic process and reduce the testing burden by being used as a pre-screening for biopsies. The application could improve everyday clinical practice and increase the benefits for patients. The use of transformer neural networks represents an advancement over the previously used convolutional neural networks and offers better performance and robustness.

Fig. 4.19 iPhone 4 vs. iPhone 14. Comparing next-generation sequencing (NGS) to the early generation of sequencing methods is like comparing the iPhone 4 to the iPhone 14

4.9   Cancer Grading and Staging

Medical staff classify cancer to see how severe the cancer is and what can be done about it. The terms *grading* and *staging* are often used interchangeably. However, from a medical perspective, they are different. Grades describe the appearance of cancer cells, while stages tell us the size and spread of the tumor, determined by multiple factors. Cancer is graded based on criteria such as how strongly the tumor cell differentiates from its healthy environment. If the cancer cells look like normal healthy cells, then the malignancy is thought to be low. However, a highly malignant cancer cell is less differentiated and has lost the appearance and characteristics of a healthy cell. Malignancy ranges from Grade 1, where cancer cells are highly differentiated and not highly malignant, to Grade 4, where cells are not differentiated and highly malignant.

Cancer stages tell us about the level of invasion by cancer. Stages tell us if the cancer is local or regional or if it has spread to other parts of the body. Staging is a cancer-specific classification. There are four stages for most types of cancers, ranging from stage I to stage IV. Stage I signifies the least threatening stage of cancer, while stage IV means the cancer is in its most aggressive form. The **TNM system** determines the stages. "T" is tumor size, "N" is the number of lymph nodes that have been infiltrated, and "M" means whether metastasis has occurred. TNM staging increases in severity with numbers. A T2 tumor is bigger than a T1 tumor. An N3 tumor has more lymph nodes involved than an N1 tumor. Furthermore, M0 means the tumor has not metastasized to other organs, while M1 means it has. For most types of cancer, if a tumor has metastasized (i.e., M1), it is stage IV cancer. Metastasis to other organs is the hallmark of aggressively malignant tumors.

Knowing what stage the cancer is in helps doctors determine which treatment options are possible and what the rates of survival are. People with stage IV cancer have a significantly lower chance of survival than people with stage I cancer. For stage IV cancers, palliative therapy may be more suitable than curative therapy.

4.10 Excurse AlphaMissense (Innovation)

AlphaMissense Revolutionary AI in biotechnology: Biotechnology is on the verge of a groundbreaking change, thanks to a new artificial intelligence (AI) called AlphaMissense [20], developed by DeepMind. This isn't just another AI; it's a revolutionary technology that uses extensive genetic databases and deep learning to classify missense mutations in the human genome with unprecedented accuracy and scalability. Missense variants are specific DNA errors that can lead to altered proteins and potentially cause diseases such as cancer. The potential of this technology is truly exciting.

AlphaMissense: superior variant scoring through AI and biology. This innovative AI has already identified 71 million of the more than 216 million possible point mutations in the human genome as missense mutations [21]. Crucially, it has classified 32% of these as potentially pathogenic and 57% as probably harmless. This classification is not just theoretical, as it has real-world applications in the diagnosis of rare diseases and biotechnological processes such as protein design. By combining a mutation pattern recognizer, a language model for DNA base codes, and a variant of AlphaFold that predicts protein folds, AlphaMissense can accurately predict whether a missense mutation is harmful or harmless.

AlphaMissense: impressive performance and far-reaching implications. This AI is not just a theoretical advancement—it significantly outperforms previous methods in classifying variants. It can classify 92.9% of variants as benign or malignant with 90% accuracy, compared to 67% of earlier methods. This precision and comprehensiveness have tangible benefits for biotech startups, enabling them to avoid risky genetic experiments and identify promising candidates for laboratory testing. This, in turn, accelerates the development of personalized gene therapies, optimized enzymes, and biopesticides, potentially revolutionizing the biotechnology industry.

New horizons in genetic engineering: For startups looking to push the boundaries of biological reprogramming, AlphaMissense mitigates technological risk. Whether creating new protein scaffolds or engineering microbial strains, the technology reduces the lengthy and expensive design-build-test cycle. By rapidly filtering billions of genetic variants for promising candidates worthy of laboratory validation, AlphaMissense significantly increases the chances of success.

Accelerating the synthetic biology revolution: The industrialization of biology requires reliable predictions of how genetic changes will affect phenotypic changes before costly testing is performed—a problem that AlphaMissense essentially solves. Through program-driven optimization of protein function, AlphaMissense advances the critical goals of synthetic biology, such as greater efficiency in design-build-test cycles and complexity management. This paves the way for cloud platform startups to offer AlphaMissense-as-a-Service for enterprise biotechnology, greatly expanding opportunities and fostering a sense of optimism for startups combining recent advances in genomics and machine learning.

The future of biotechnology: AlphaMissense marks the beginning of an exciting era in biotechnology. Startups can develop breakthrough applications faster and cheaper without genetic uncertainties. This technology could spark a wave of entrepreneurship at the intersection of AI and biology and drive the industrialization of biology. While innovations like AlphaFold democratized access to protein structure knowledge, AlphaMissense makes what was previously considered genomic "dark matter" accessible and enables exploration into unknown territories. The industrial revolution of the twenty-first century could thus be at the heart of bioengineering.

Innovative approaches to artificial intelligence will play a decisive role in the future. Google DeepMind has developed AlphaMissense, a software specifically designed to recognize and evaluate missense mutations in DNA. This

AI uses AlphaFold information to deduce a protein's three-dimensional structure from its genetic sequence. Although AlphaMissense cannot predict the exact changes in protein structure caused by a mutation, it helps to predict the likelihood of a disease. This technology expands our knowledge of human genes and offers promising opportunities to advance genetics research and improve medical diagnostics.

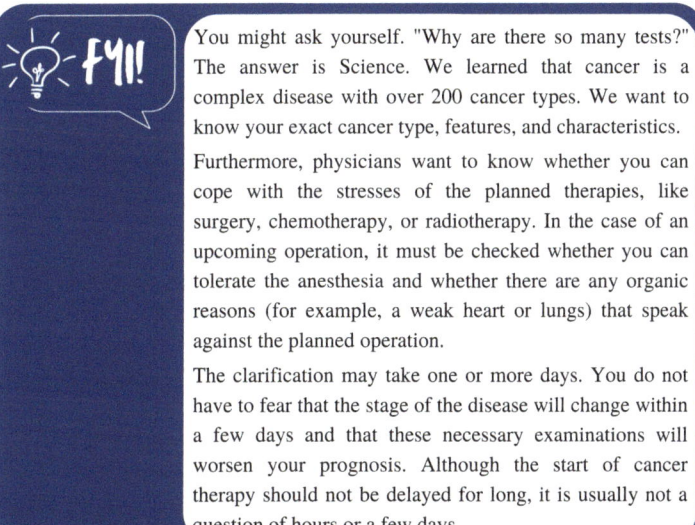

You might ask yourself, "Why are there so many tests?" The answer is Science. We learned that cancer is a complex disease with over 200 cancer types. We want to know your exact cancer type, features, and characteristics.

Furthermore, physicians want to know whether you can cope with the stresses of the planned therapies, like surgery, chemotherapy, or radiotherapy. In the case of an upcoming operation, it must be checked whether you can tolerate the anesthesia and whether there are any organic reasons (for example, a weak heart or lungs) that speak against the planned operation.

The clarification may take one or more days. You do not have to fear that the stage of the disease will change within a few days and that these necessary examinations will worsen your prognosis. Although the start of cancer therapy should not be delayed for long, it is usually not a question of hours or a few days.

Summary
In this chapter, we have gained a deep understanding of the methods of cancer diagnosis. We have learned about the various techniques and examinations, such as medical history, physical examinations, blood tests, imaging techniques, tissue samples, and genetic analysis used by medical professionals to diagnose cancer. We have highlighted the integration of artificial intelligence in medical diagnostics, the relevance of genetic tests, the development of liquid biopsies, and the use of AI in medical diagnostics. This chapter provides a comprehensive account of the different approaches and technologies used to diagnose cancer. We have also discussed the vital role of artificial intelligence and digital technologies in improving the early detection and treatment of cancer. It is our collective responsibility to contribute to the battle against cancer by understanding these methods and their significance.

References

1. Kalantar-Zadeh K, et al. A human pilot trial of ingestible electronic capsules sensing different gases in the gut. Nat Electron. 2018;1(1):79–87. https://doi.org/10.1038/s41928-017-0004-x.
2. Winn Army Community Hospital > Health Services > Lab Tests & Radiology > Radiology. https://winn.tricare.mil/Health-Services/Lab-Tests-Radiology/Radiology. Accessed 05 July 2024.
3. mdr.de. Früher Tumore entdecken mit KI-Assistenten beim Ultraschall | MDR. DE. https://www.mdr.de/wissen/ki-assistenz-ultraschall-mammographie-100.html. Accessed 04 July 2024.
4. Leberkrebs besonders gefährlich—Forscher wollen ihn mit Ultraschall besiegen—FOCUS online. https://www.focus.de/wissen/natur/wissenschaft-leberkrebs-ist-besonders-gefaehrlich-forscher-wollen-ihn-mit-ultraschall-besiegen_id_236182024.html. Accessed 04 July 2024.
5. Forschende testen weltweit neue Strahlentherapie gegen Krebs. https://www.forschung-und-lehre.de/forschung/forschende-testen-weltweit-neue-strahlentherapie-gegen-krebs-6158. Accessed 04 July 2024.
6. ReporterFollow |Contact, B. B. N. L. First in Canada imaging technology means less stress, less radiation for patients at St. Joseph's. London. https://london.ctvnews.ca/first-in-canada-imaging-technology-means-less-stress-less-radiation-for-patients-at-st-joseph-s-1.6908964. Accessed 04 July 2024.
7. Faster, more precise, less radiation: State-of-the-art whole-body PET-CT officially inaugurated at UKL. https://www.uniklinikum-leipzig.de/presse/Seiten/Pressemitteilung_7706.aspx. Accessed 04 July 2024.
8. Tests and procedures used to diagnose cancer—NCI. https://www.cancer.gov/about-cancer/diagnosis-staging/diagnosis. Accessed 05 July 2024.
9. Artificial Intelligence gives cancer research a boost. 2019. https://www.youtube.com/watch?v=vhUu5vwYUak. Accessed 05 July 2024.
10. Heikenwälder H, Heikenwälder M. Der moderne Krebs—Lifestyle und Umweltfaktoren als Risiko. Berlin/Heidelberg: Springer Berlin Heidelberg; 2023. https://doi.org/10.1007/978-3-662-66576-3.
11. Genomic vs. genetic testing for cancer. https://www.cancercenter.com/diagnosing-cancer/genetic-and-genomic-testing. Accessed 05 July 2024.
12. Hu L, et al. Fluorescence in situ hybridization (FISH): an increasingly demanded tool for biomarker research and personalized medicine. Biomark Res. 2014;2(1):3. https://doi.org/10.1186/2050-7771-2-3.
13. Vela V, Juskevicius D, Gerlach MM, Meyer P, Graber A, Cathomas G, Dirnhofer S, Tzankov A. High throughput sequencing reveals high specificity of TNFAIP3 mutations in ocular adnexal marginal zone B-cell lymphomas. Hematol Oncol. 2020;38(3):284–92. https://doi.org/10.1002/hon.2718.

14. Vela V, Juskevicius D, Prince SS, Cathomas G, Dertinger S, Diebold J, Bubendorf L, Horcic M, Singer G, Zettl A, Dirnhofer S, Tzankov A, Menter T. Deciphering the genetic landscape of pulmonary lymphomas. Mod Pathol Off J U S Can Acad Pathol Inc. 2021;34(2):371–9. https://doi.org/10.1038/s41379-020-00660-2.
15. Risk factors for chronic myeloid leukemia. https://www.cancer.org/cancer/types/chronic-myeloid-leukemia/causes-risks-prevention/risk-factors.html. Accessed 05 July 2024.
16. Molekularbiologische Methoden. https://www.krebsinformationsdienst.de/untersuchungen-bei-krebs/molekulare-diagnostik/methoden. Accessed 05 July 2024.
17. Ärzteblatt, D. Ä. G., Redaktion Deutsches. Das Pankreaskarzinom auf dem Weg zur Präzisionsmedizin. Deutsches Ärzteblatt. https://www.aerzteblatt.de/nachrichten/149525/Das-Pankreaskarzinom-auf-dem-Weg-zur-Praezisionsmedizin. Accessed 15 July 2024.
18. Sosinsky A, et al. Insights for precision oncology from the integration of genomic and clinical data of 13,880 tumors from the 100,000 genomes cancer programme. Nat Med. 2024;30(1):279–89. https://doi.org/10.1038/s41591-023-02682-0.
19. Wagner SJ, et al. Transformer-based biomarker prediction from colorectal cancer histology: a large-scale multicentric study. Cancer Cell. 2023;41(9):1650–1661.e4. https://doi.org/10.1016/j.ccell.2023.08.002.
20. Podbregar N. KI entschlüsselt unsere Mutationen. scinexx | Das Wissensmagazin. https://www.scinexx.de/news/technik/ki-entschluesselt-unsere-mutationen/. Accessed 04 July 2024.
21. Shibil. AlphaMissense: how this AI breakthrough will revolutionize biotech ventures. Medium. https://medium.com/@profshibil/alphamissense-how-this-ai-breakthrough-will-revolutionize-biotech-ventures-fe8e02017c8f. Accessed 04 July 2024.

5

The Psychological Impact of Cancer

Contents

Abstract Cancer is a formidable illness that profoundly impacts us physically and emotionally. Receiving a cancer diagnosis can be incredibly jarring, evoking feelings of fear, stress, and sorrow. Nevertheless, there is reason for optimism, as advancements in diagnostics and treatments have greatly improved the prognosis for many types of cancer. Some are now curable, while others can be effectively managed. This chapter will delve into the emotional and practical hurdles faced by cancer patients and their loved ones and offer guidance on navigating this challenging period with the support of family, friends, and self-help groups.

The good news nowadays is that cancer is no longer always a death sentence. Cancer prognosis has improved significantly recently. Thanks to modern therapies, some types of cancer are now curable, and many are treatable and even controllable.

Cancer is different from other diseases you deal with all the time because it often changes you both physically and emotionally. Upon hearing the word *cancer*, many people experience shock. They become paralyzed with fear and panic. Before this life-changing moment occurred, you thought you had plenty of time to live ahead of you, yet suddenly, you may not. This scenario is understandably stressful.

Then, there is a period of uncertainty while you wait for a definite diagnosis. This period can take anywhere from several days to more than a week or so. If you discover that you have cancer, this is when your perspective changes, and you feel a sense of urgency. Everyone who has just been diagnosed with cancer realizes it may threaten their life. Learning more about the disease can help us understand what further examinations can be done and what therapies there are. Having this information can help reduce our fear and help us face the disease with more confidence. Nevertheless, the good thing is that we are not alone. During this time, it is helpful to communicate with trusted friends, family members, medical staff, and even patient associations to help us cope with physical and emotional symptoms.

5.1   Anxiety

At different times during the treatment, it is common to have anxiety. While cancer therapy may last a long time, anxiety is one way we respond to emotionally stressful situations. Our sense of ***fear, worry, and uneasiness*** can trigger anxiety [1]. A cancer diagnosis can make us afraid of "change" or "losing control." All of a sudden, we face uncertainty, the unknown and possibly death. We begin to ask what will happen next. Will I have to cancel the holidays? Stop working? Lose my job? Lose my friends? Feel older? Have a dangerous surgery? Risk dying? It is normal to ask these questions. In some cases, anxiety can also cause sudden and acute panic attacks, which occur suddenly with or without warning, often reaching their worst point within 10 min. If you feel anxiety symptoms most of the day, nearly every day, or if they interfere with your daily life, you should contact a health professional.

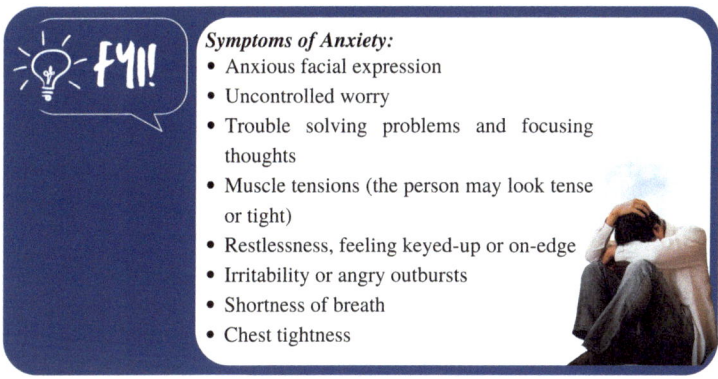

Symptoms of Anxiety:
- Anxious facial expression
- Uncontrolled worry
- Trouble solving problems and focusing thoughts
- Muscle tensions (the person may look tense or tight)
- Restlessness, feeling keyed-up or on-edge
- Irritability or angry outbursts
- Shortness of breath
- Chest tightness

5.2   Stress

When we are threatened, our body releases hormones like adrenaline. Such hormones increase heart rate and blood flow to the legs and decrease our reaction time. This is known as **the fight-or-flight response** we experience in stress. Once the threat eases, our body relaxes and our hormone levels also decline. The initial shock that comes from the diagnosis that you have cancer is one reason why it is **emotionally taxing.** We also face long-term stress since we have to face daily life with the disease. Therefore, cancer produces both short-term and long-term stress in patients. During chronic stress, we have high hormone levels, which can damage our bodies and immune systems [2] even further. Therefore, these harmful biological effects may diminish our ability to fight cancer since our immune system, which is already under pressure, may not function correctly.

Chronic stress significantly impacts the immune system, accelerates the aging process, and can even shorten life expectancy. Studies have found that stressful life events, like the loss of a loved one or a job, can increase the risk of certain cancers. For instance, men who experience high levels of job-related stress have a markedly higher risk of developing prostate cancer [2].

Stress hormones like cortisol, adrenaline, and noradrenaline play crucial roles in these processes. They prepare the body for action, boosting the performance of muscles, the heart, and the lungs. However, cortisol, in particular, has long-lasting effects and can suppress immune cell activity during prolonged stress, leading to a higher susceptibility to infections and inflammation—both of which can elevate cancer risk [3]. Additionally, social stress can lead to excessive immune system activation and chronic inflammation [2], a well-known risk factor for cancer. Psychology refers to the capacity to bounce back from stress as resilience. People with low resilience to stress are at a greater risk of developing specific cancers, such as liver and lung cancer.

Other contributing factors, including smoking, an unbalanced diet, and sleep disorders, can worsen the problem. Stress often leads to sleep deprivation, which is also a significant risk factor for cancer. Research has shown that sleep disorders can increase the likelihood of breast, prostate, and bowel cancers, especially in those who work night shifts or irregular hours.

Experts suggest that exposure to light at night may disrupt melatonin production, leading to the increased cancer risk associated with sleep deprivation [2]. Melatonin is a hormone that regulates the sleep-wake cycle and also inhibits tumor cell growth. Therefore, a deficiency in melatonin could heighten the risk of hormone-dependent cancers, such as breast and prostate cancer.

In conclusion, stress has a significant impact on health and cancer risk. Prolonged stress, unhealthy lifestyle choices, and insufficient sleep can weaken the immune system, promote inflammation, and ultimately raise the risk of cancer. It is crucial to recognize stress and take proactive steps to mitigate its harmful effects. Maintaining a healthy lifestyle—including adequate sleep, a balanced diet, and regular exercise—can help reduce the risk of cancer and enhance overall well-being.

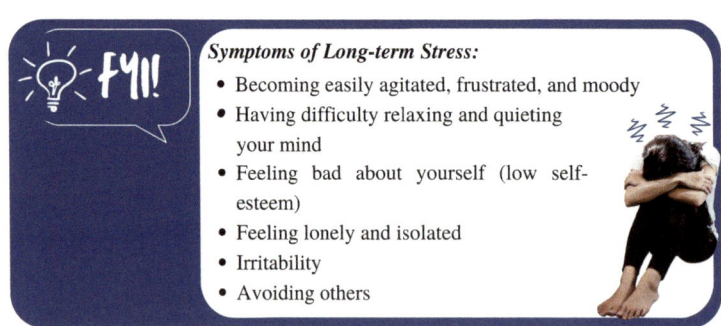

Symptoms of Long-term Stress:
- Becoming easily agitated, frustrated, and moody
- Having difficulty relaxing and quieting your mind
- Feeling bad about yourself (low self-esteem)
- Feeling lonely and isolated
- Irritability
- Avoiding others

5.3   Depression

It's normal to feel sad when faced with cancer. Sadness accompanies losing something, whether that something is health or the freedom to do certain activities (travel, sport). For example, it may also mean giving up specific projects. Depression is more than just sadness. Depression is associated with a feeling of hopelessness and tends to "paralyze" the patient, who finds it challenging to find the strength to do anything and thus get out of this state. Despair makes it challenging to make decisions (e.g., about treatment) or to follow a prescribed treatment plan. Depression can be connected to feeling tired, which can make feelings of treatment fatigue even worse. It can also be related to difficulties in eating properly, which can increase the feeling of fatigue.

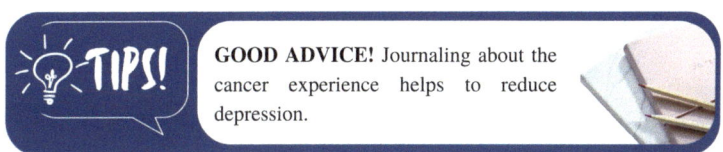

GOOD ADVICE! Journaling about the cancer experience helps to reduce depression.

Elevated levels of inflammatory markers like interleukin-6, CRP, and TNF-α have been detected in healthy individuals who experience depression or are prone to aggression. Although these two conditions exhibit different behaviors, they are both linked to stress, which is believed to drive the increased production of these inflammatory substances. This heightened inflammation may contribute to the development of both depression and aggression [3].

There is a possibility that stress, depression, and inflammation in chronic diseases may interact and worsen each other. Following a cancer diagnosis, many patients experience depression, which is often associated with a poorer prognosis [4]. These depressive symptoms may not only stem from the diagnosis itself but could also be influenced by the inflammatory processes that are common in nearly all cancers [5]. The inflammatory substances might impact mental health in a manner similar to clinical depression, suggesting that depression in cancer patients could be a sign of increased inflammatory activity within the tumor environment, potentially accelerating cancer progression.

Scientific research indicates that stress-induced inflammatory responses may play a role in the onset of depressive disorders. In the future, certain patients with depression—especially those with elevated inflammation levels and resistance to traditional treatments—might benefit from anti-inflammatory medications.

Interestingly, certain viral infections, like those caused by herpes viruses (e.g., varicella-zoster, herpes simplex virus type 2) and influenza, are also considered potential risk factors for depression [6]. In these cases, the body's inflammatory response to combat the infection could directly trigger depressive symptoms or worsen preexisting depression.

Furthermore, postinfectious depression, which can emerge weeks to months after an influenza infection and presents with depressive symptoms, is also linked to the immune system's inflammatory response. Despite its prevalence, this form of depression remains relatively unknown to many people.

Symptoms of Clinical Depression:
- Ongoing sad, hopeless or "empty mood" for most of the day
- Loss of interest or pleasure in almost all activities, most of the time
- Feeling slowed down or restless and agitated almost every day, enough for others to notice
- Extreme tiredness (fatigue) or loss of energy
- Trouble sleeping with early waking, sleeping too much or not being able to sleep
- Trouble focusing thoughts, remembering or making decisions
- Feeling guilty, worthless or helpless

5.4   Dealing with Family and Friends

Only you have the authority to choose when and how you tell others you have cancer. Someone may have gone through the diagnostic process with you, but there are also family members or friends you will need to tell. Remember that only you have the power to tell whom you want, what you want, and when. *Do it in your time.* Telling your parents might be the most difficult, but they are there for you and will help you through this challenging period.

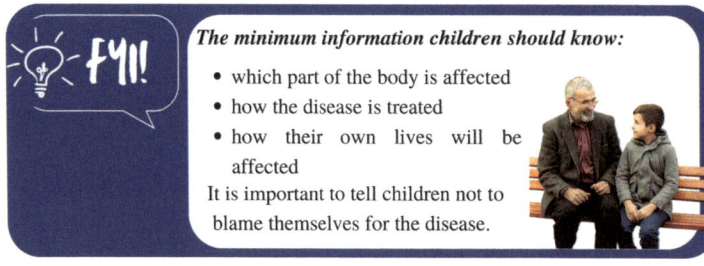

The minimum information children should know:
- which part of the body is affected
- how the disease is treated
- how their own lives will be affected

It is important to tell children not to blame themselves for the disease.

Telling close family members that you have cancer is very hard, but it is even more difficult when you are the parent of young children since they are very sensitive. They tend to pick up on far more cues and information than they are given credit for. Children may suspect the worst if they are not confronted with honest, accurate information. To a child, the most important thing to say is that they are not responsible for the disease. When you talk to your young children (8–12 years), they will not need much detailed information, but teens and older children need to know more. Some hospitals and cancer centers also have support groups for children whose parents have cancer.

5.5   The Role of Faith

While faith and medicine do not always go hand in hand, we cannot deny the impact faith has on the lives of cancer survivors. Faith has helped cancer patients deal with the uncertainty they are facing due to the disease. *Prayer* is often seen as a powerful coping mechanism. Simply asking for the presence of God on the journey could help strengthen patients and reduce their stress. Those who pray regularly are less likely to suffer from depression and anxiety. Increasing satisfaction in life can also decrease pain. *Meditation* may also achieve these effects for those who do not believe in a deity.

5.6   Support Groups

Recently, the network of support groups has notably grown. Today, there are many ways to find support groups. You have the option to choose from groups you *attend in person* or online in *moderated chat rooms*. There are groups run by *cancer organizations* that provide support for virtually every type of cancer. Furthermore, there are also groups for cancer survivors. In short, there are many organizations. Some offer education, some focus more on recreation and socializing, while others are formed for caregivers. Many people report that they find tremendous emotional support in such groups, while others are less enthusiastic. However, the choice is yours. If you do not feel comfortable attending such groups or social spaces, consider searching for other options. However, if you want to visit a group, the exploration may result in a worthwhile journey.

5.7 Find a Mentor

You can also find support by having a mentor. Once you are diagnosed with cancer, the desire to find a person who can understand and support you is normal. Friends are some of the best sources of support, but they may not understand much about cancer. Nowadays, organizations like "Mentor Angels," match cancer patients with survivors according to age, gender, and type of cancer. As one organization notes, "A Mentor Angel is a walking, talking, living proof and inspiration that cancer can be beaten." Many cancer survivors also find volunteering for such an organization helps them immeasurably. In sum, as this chapter shows, a cancer patient should never feel alone, with nobody to talk to or help them.

In this chapter, we have highlighted the emotional and psychological challenges of a cancer diagnosis. This chapter emphasizes that cancer is no longer necessarily a death sentence but that modern therapies have improved the prognosis for many cancers. Nevertheless, the diagnosis often causes anxiety, stress, and depression as we are faced with uncertainty and changes in our lives. In this chapter, we have learned the importance of support from our family, friends, medical staff, and especially support groups. These groups, filled with individuals who understand our struggles, can provide a sense of belonging and comfort. In addition, faith plays an important role, and the opportunity to find a mentor who can be a source of inspiration and comfort based on their own experience with cancer is also significant. The message is that as cancer patients, we are not alone and can use various resources to cope with our physical and emotional symptoms.

CAUTION!

Remember, a positive attitude is not a cure for cancer. The immune system cannot magically fight cancer better because you are relaxed. However, focusing on the positive things can provide the mental and emotional strength to deal with adverse circumstances such as cancer. This helps the healing process move forward.

References

1. Libov C. Cancer survival guide: how to conquer this disease and live a good life. 1st ed. Humanix Books; 2016.
2. Heikenwälder H, Heikenwälder M. Der moderne Krebs—Lifestyle und Umweltfaktoren als Risiko. Berlin/Heidelberg: Springer Berlin Heidelberg; 2023. https://doi.org/10.1007/978-3-662-66576-3.
3. Takahashi A, Flanigan ME, McEwen BS, Russo SJ. Aggression, social stress, and the immune system in humans and animal models. Front Behav Neurosci. 2018;12:56. https://doi.org/10.3389/fnbeh.2018.00056.
4. Sotelo JL, Musselman D, Nemeroff C. The biology of depression in cancer and the relationship between depression and cancer progression. Int Rev Psychiatry Abingdon Engl. 2014;26(1):16–30. https://doi.org/10.3109/0954026 1.2013.875891.
5. Grivennikov SI, Greten FR, Karin M. Immunity, inflammation, and cancer. Cell. 2010;140(6):883–99. https://doi.org/10.1016/j.cell.2010.01.025.
6. Gale SD, Berrett AN, Erickson LD, Brown BL, Hedges DW. Association between virus exposure and depression in US adults. Psychiatry Res. 2018;261:73–9. https://doi.org/10.1016/j.psychres.2017.12.037.

6

Cancer Therapies

Contents

Abstract This chapter explores the diverse landscape of cancer treatments, emphasizing the need for personalized care. It covers local treatments like surgery and radiotherapy, systemic approaches such as chemotherapy and hormone therapy, and cutting-edge therapies like CAR-T and CRISPR-CAS9, offering hope for complex cases. Treatment decisions are based on factors such as cancer type, patient age, and relapse potential, with multidisciplinary teams collaborating through tumor boards to provide tailored options. Patients are encouraged to discuss all treatment paths, including palliative care, with their doctors. For some, a wait-and-see approach may be suitable for slow-growing cancers, while others may benefit from early, aggressive intervention. Advances in modern technology and targeted therapies give patients more precise, effective treatment choices, boosting confidence in their care. The chapter also stresses the importance of mental health support and integrative medicine to help patients manage the emotional suffering and potential fatigue during their cancer journey.

6.1   Overview of Cancer Treatment

After a cancer diagnosis, the next goal is to find the best treatment. To accomplish this, doctors from different specialties meet and participate in multidisciplinary team meetings, known as a ***tumor board.*** Each board member puts forward their points of view to make decisions together based on information gathered from diagnostic tests, including scientific data, their experiences, and available clinical trials. The specialists consider whether surgery is a good idea, what type of therapy is most efficient, and if the person can endure possible side effects. However, a patient can always choose to "let nature drive the course of cancer" and ask only for palliative care.

Next to a medical meeting, the chosen proposal is discussed with the patient. The patient's referring doctor then takes responsibility for translating the proposal into simple language and providing necessary explanations to help the patient validate or reject the specialist's proposal. The patient's decision is the final one, even in severe illnesses. As long as the patient fully understands the situation, they have the right to choose whether to seek treatment or let nature take its course. The doctor's crucial role is to provide patients with all the information they need to make the right decision and support them in their choices for the rest of their lives. Remember, you are in charge of your health, and your doctor is there to help you make informed decisions that are best for you.

But if treatment is the choice, the ultimate goal is to find the optimal way to fight cancer while minimizing harm to the patient. By working together, doctors can ensure that every person receives the personalized care they require. One of the most challenging things about battling cancer is that there is ***no one-size-fits-all solution***. The disease is incredibly diverse, which is why different patients will respond to treatments differently. As a result, oncologists (cancer specialists) must carefully consider each case individually before deciding with the tumor board which course of treatment to follow. In some cases, local therapies like surgery or radiotherapy may be effective. However, more aggressive measures like chemotherapy may be necessary in other cases. The key is to tailor the treatment to the patient's individual needs. Only by doing this can we hope to achieve the best possible outcome for an individual (Fig. 6.1).

The patient's *age* plays an essential role in determining whether to undergo cancer treatment. While a younger patient may be more likely to tolerate aggressive treatment and have a longer life expectancy, an older person may not be able to handle the side effects of treatment physically or psychologically. In addition, fertility preservation measures may need to be taken for

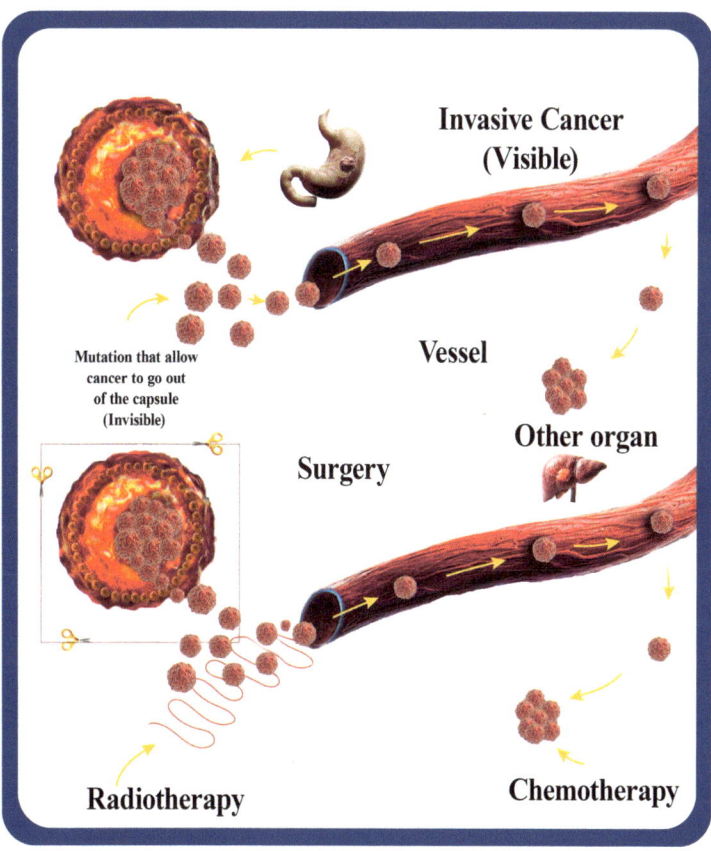

Fig. 6.1 Choosing the treatment type. If the tumor is locally invasive and has not metastasized, local treatments such as surgery or radiotherapy are possible. On the other hand, if the cancer is aggressive and metastatic, chemotherapy might be mandated to help kill the cancer cells that are on their way to invade other organs

younger people undergoing treatments that could affect their ability to have children in the future. Patients' ***overall health*** must be considered when making treatment decisions, as well as their personal situation (e.g., Do they have support from family and friends?). Considering all aspects is key to making the best decision.

It is crucial to be aware that proposing specific restrictive treatments to patients with different needs and values can be equally hazardous. For instance, prescribing a rigorous treatment plan to a young drug-addicted patient who fails to attend appointments regularly or to an older adult who treasures their independence but doesn't consume adequate food or fluids can be precarious. Therefore, it is imperative to consider individual requirements and preferences while devising treatment plans.

Of course, before making any decisions, weighing the pros and cons is always essential. This is especially true when it comes to medical treatment. Before deciding on a course of action, patients should discuss all options with their doctor, including the goals and duration of the treatment, as well as any possible side effects. Furthermore, we should also consider the ***convenience*** of the treatment, including logistical issues like how far the medical facility is from our home. If the inconvenience of the treatment outweighs the potential benefits, we have the right to refuse the treatment a doctor proposes. By fully understanding all aspects of the situation, we can make informed decisions about our medical care.

Taking a break before starting a therapeutic project can be a wise decision. It allows patients to complete important tasks before treatment begins, which may take a long time. With certain types of slow-growing cancers like low-grade lymphomas and some prostate cancers, a "watch-and-wait" approach may be beneficial. By monitoring the tumor growth, doctors can avoid unnecessary treatments and unwanted side effects that can affect the patient's quality of life. Instead of rushing into treatment, doctors can wait for the cancer to grow to a stage where it is necessary to treat, which could take years. This approach can help patients avoid the stress of rushing into a treatment plan and focus on their personal goals before starting a therapeutic journey.

In some cases, low-grade lymphomas will even go into partial remission while waiting. This strategy is also used for elderly people with prostate cancer. They might not want to have treatment sessions every day. Some elderly patients die of other causes before they even need cancer treatment. However, it is important to remember that strict parameters are followed and checked regularly during the wait. As soon as the need arises, treatments will start. When we fully understand all of our options, we can make informed decisions about our medical care.

Usually, the most effective treatments are those given first. The "second choice" of treatment would normally only be given if there is a recurrence. If this second-choice treatment does not help, a "third" choice might be available. For example, doctors may ask patients if they want to take part in a clinical trial that tests a novel treatment. This option will provide new avenues of cancer research to find more effective treatments.

Of course, we may also enter a ***clinical trial*** right from the start when our cancer treatment begins. Yet, entering a clinical trial can be a tough decision to make. On the one hand, you are investing your time and energy into something that may or may not work. On the other hand, you are giving yourself a chance to receive treatments that could save your life. It is a personal decision that each of us has to make for ourselves. However, if you decide that a clinical trial is right for you, know that you are not alone. Many people have taken the brave route of going there before you, and many more will come afterward, too. You are part of a long line of people fighting for their lives and doing whatever it takes to beat cancer.

6.2 Local Treatments

Local treatment is a type of cancer therapy that involves ***targeting the tumor*** with surgery or radiation while sparing the surrounding healthy tissue. Local treatment is often used when the ***cancer is confined to a specific area*** and has not spread to other body parts. Local treatment can effectively control cancer and improve the quality of our life. It is even possible to cure cancer with local treatment alone. However, local treatment is not always appropriate in some cases, particularly for cancers that have a high potential to spread rapidly to other parts of the body. In such cases, chemotherapy may be the most effective option, as is the case with high-grade lymphomas, a type of white blood cell cancer. By avoiding local treatment in these instances, cancer progression can be more effectively managed. That is why we should discuss our options with our doctor before making any final decisions. We have the opportunity to discuss all of your questions regarding the best course of action, including the possibility of surgery, radiotherapy, and any other available options. A team of specialists from each respective field will be present at the meeting to provide their expert opinions and guidance.

6.3   Surgery

Most people consider surgery to be a last resort. However, surgery is often the best chance for a cure, especially if cancer has not spread to other organs. Therefore, surgery usually works best for *solid tumors.* Usually, cancer of the blood, like lymphomas, leukemias and myelomas, are not treated by surgery. This is because if the primary tumor is removed, the migrating cancer cells are not treated. Instead, a combination of treatments, such as radiotherapy for the localized cancer cells and drugs circulating in the bloodstream to treat the migrating cancer cells, is preferred.

In general, cancers with a high risk of metastasis, composed of cells that have a strong tendency to migrate into blood vessels, such as small-cell lung cancers, are not treated with simple surgical procedures, even if they are localized. Indeed, operating on the main tumor has no impact on the migrating cancer cells that later form visible metastases. It is these migrating cells that ultimately affect a patient's prognosis. Imagine a wasp nest in your house: if you remove it and most of the wasps have left, removing it will not prevent the wasps (who have left) from building another one elsewhere in your house. Similarly, destroying the main tumor through surgery or radiotherapy won't suffice to treat the cancer if cancer cells have spread to other organs.

There are some cases where metastases may still be treated with surgery despite being uncommon. If the metastases are limited in number and progressing slowly, or if they are causing significant discomfort to the patient, surgery may be a viable option.

Additionally, in life-threatening situations where the cancer has already metastasized, surgery may be performed as a means of relieving the patient's condition. For example, a patient with a cancerous tumor blocking their digestive tract may require surgery to remove the tumor and to eliminate any migrating cells, even if it has already spread to other areas, as the blockage would be life-threatening if left untreated.

If surgery cannot eliminate all cancer, it can still be an essential part of any treatment. Cancer surgery has seen remarkable advancements in recent years thanks to cutting-edge technology. "Conservative" surgery for breast cancer has eliminated the need for complete breast removal in every cancer case, provided surgery is accompanied by radiotherapy. Patients can now enjoy a faster recovery time through the use of minimally invasive surgery, while robot-assisted surgery offers greater precision to minimize nerve damage during prostate surgery. However, these techniques may not be ideal or applicable in all cases. In situations where there is a high hereditary risk, it may be necessary to remove the entire breast instead of only the cancer. This helps to mitigate the risk of relapse or new tumors. With these advancements, cancer patients can now benefit from more personalized and less invasive surgical options, leading to a better quality of life.

Additionally, minimally invasive surgery allows for faster recovery, while robot-assisted surgery offers greater precision, which is particularly crucial for sparing delicate nerves in procedures like prostate surgery.

However, these techniques aren't universally applicable or desirable. In some cases, it may still be necessary to remove the entire breast rather than just the cancer, especially in instances of hereditary risk.

Cancer surgery is a standard treatment that aims to remove the tumor completely, along with some of the surrounding tissue and any lymph nodes that might contain cancer cells as well. Getting rid of as much cancer as possible is crucial because it is easier to treat when it is still confined to one area.

When it comes to cancer surgery, there are instances where the surgeon may not be able to remove the entire tumor due to its location or size. In such cases, additional treatments can be used to help reduce the tumor's size, which could make surgery more feasible or destroy any remaining cancer cells after surgery. These treatments, known as "neoadjuvant" and "adjuvant," have shown promising results in such cases. Therefore, it is essential to consider these treatments to ensure the best possible outcome for the patient.

The edge of the tissue is called the margin. The pathologist will look at the margins under a microscope to see if all cancer has been removed. If cancer cells are not found at the margin, it is called a *negative margin* (Fig. 6.2). This means that there is a good chance that all cancer has been removed. However, if there are cancer cells at the tissue margin, then it is called a *positive margin,* which means there is a chance that not all cancer has been removed. Sometimes, not every single cell can be removed with one surgery, so that another surgery may be needed.

In some cases, even if there is a positive margin, there may not be any recurrence. However, the surgeon usually prefers to perform a re-intervention to ensure that no cancer cells are left locally, which could potentially cause a recurrence. In situations where a second operation is impossible, radiotherapy is employed to complete the treatment. Typically, tumors appear as a mass with small "roots" that are not visible to the surgeon and can only be detected when the cancer is analyzed under a microscope (Fig. 6.3). In such cases, the margin that cuts the roots is referred to as "positive." The type of surgery performed depends on so many factors, such as your overall health and the size and location of your tumor. In most cases, surgery successfully removes all the cancerous tissue. Hence, the margin is "negative." However, there is always a

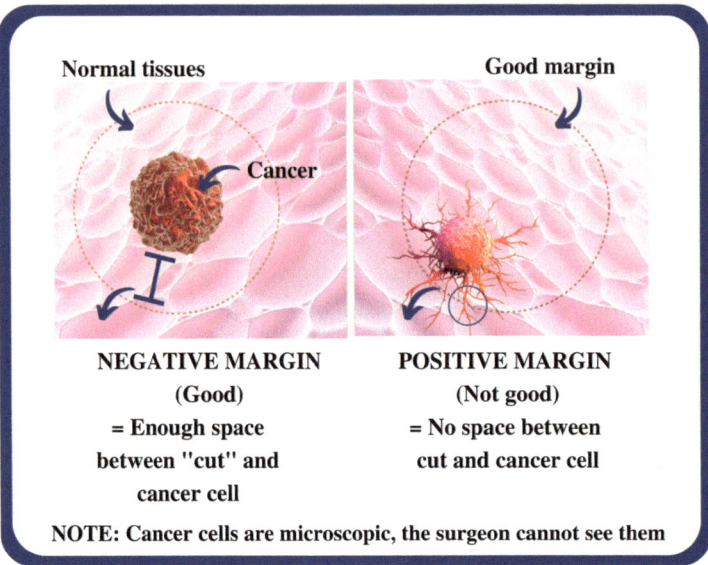

Fig. 6.2 Surgical Margins. During surgery, the tumor margins can be classified as negative, where the entire tumor is removed, or positive, where new interventions might be needed

Fig. 6.3 The picture is an illustration showing two lawnmowers on a grassy area with two tumors. The lawnmowers represent treatments used to mow the grass, or in this case the tumor. The cancer cells beneath the surface represent the hidden or underlying problem that the lawnmowers (treatments) are intended to address. This visual metaphor suggests the concept of tackling cancer at the surface as well as at the root and illustrates how difficult it is to eradicate tumors completely

risk that some cancer cells might remain after the surgery. Hence, the margin is "positive."

Oncologists often tell patients that the surgeons cannot see the cells they operate on. The cells are not visible to the human eye; they can only be seen with a microscope. The surgeon does not operate under a microscope.

However, for certain delicate operations, a specialist called a pathologist may attend the operation and analyze the microscope samples as they are taken. In the field of dermatology, the surgeon himself sometimes alternates between surgery and observation under the microscope. For example, a type of surgery called MOHS is performed on facial tumors to keep scars as small as possible.

Only then can we better understand the situation. Looking at a tumor is like looking at a city from an airplane: you can see the city as a whole but not individual people so easily. To observe individual people from a plane, it is necessary to get closer to the ground. The microscope enables you to do exactly that, allowing you to closely observe the cells individually (Fig. 6.4) and analyze their margins with precision. Unfortunately, if cancer cells remain in the body after surgery, they can sometimes grow and form new tumors. That is why we need to continue to receive ***follow-up care*** after surgery.

While surgery is still the most common way to treat cancer, some types of cancer cannot be treated with surgery. This is usually because the cancer is located in an unfavorable position or has already metastasized. Even if all the metastases were surgically removed, invisible cancer cells could still be present in the blood vessels. This is why surgery is ineffective for liquid cancers such

Fig. 6.4 View from a plane. Cancer cells are not visible to the naked eye, just like how humans on the ground are not visible from a flying plane

as leukemia or lymphoma. Instead, these types of cancer are treated with drug therapies and sometimes with radiation therapies.

It is important to carefully consider the risks and benefits of any surgical procedure, especially when it comes to tumors. The risks associated with the tumor depend on various factors like its size and location, which the surgeon will assess. If removing a large liver tumor left insufficient "healthy" liver to function correctly after the operation, other options should be considered. Similarly, if there are life-threatening risks or severe complications with removing a tumor that is "stuck" to the heart or close to vital structures in the brain, other options should be considered. Apart from the risks associated with the tumor, there are also risks associated with the use of anesthesia. The anesthetist assesses these risks by considering the patient's age, smoking habits, preexisting medical conditions, and certain medications. They have scores to pinpoint these risks and adjust the anesthesia dosage accordingly to minimize complications.

In some cases, the risks may be so high that surgery is not recommended. This may happen with patients who already have heart disease or a strong bleeding tendency, for example. However, while it is essential to discuss all options with your surgeon before deciding, the potential benefits of surgery frequently outweigh the risk of danger [1].

Recovering from cancer surgery requires strength and resilience. Although the operation signals the end of cancer treatment, it is not the end of the journey. After surgery, patients undergo a postoperative stage critical for their recovery and well-being. In the recovery room, they receive close attention to ensure no complications arise from anesthesia. If the procedure is complex or the patient is fragile, they may require a stay in an intensive or intermediate care unit. The medical team evaluates the patient's need for pain relief, checks the scar, and ensures that the operated organ has returned to its normal state and that there is no bleeding or infection. This stage is crucial to ensure a complete and swift recovery, empowering patients to return to their daily lives as soon as possible. The patient's scar and the proper functioning of surrounding organs are then checked, and the medical team ensures that the patient is healing well. Recovery is a demanding process that requires courage and persistence. However, with the right care and support, patients can overcome the challenge and achieve a full recovery.

What should you discuss with your doctor before the operation?

- What exactly is being operated on and why?
- What type of surgery will be performed?
- How long will the surgery take?
- Are there any other treatment options?
- What happens if I do not undergo this operation?
- How should I prepare for the operation?
- Can I physically cope with the strain of the operation?
- Is the surgery ambulatory or non-ambulatory?
- What are the plans for my recovery?
- What consequences can I expect after the operation?
- Are these consequences permanent or temporary?
- What aids are available to me in case of permanent consequences?

Things to keep in mind:

- Disclose your previous illnesses and other medical problems.
- Inform your doctor of things such as smoking or drinking habits and medications taken.
- Do not hesitate to clarify any questions before giving written consent to all the procedures.
- Always consider the benefits, risks, and health consequences involved before making the decision. You can decide against specific procedures or refuse the surgery.

6.4 Radiotherapy

Radiotherapy is a local treatment that exposes cancer cells to ***high-energy rays*** that destroy their ability to reproduce. Cancer cells divide faster than most healthy body cells, which is one reason why cancer cells are more sensitive to radiation. Radiotherapy kills cancer cells or limits their growth, which can help to cure patients, lower the risk of a recurrence, and improve their quality of life. Although healthy cells are also exposed to the rays, they can repair themselves more effectively than cancer cells can. Radiotherapy is thus an effective way to specifically destroy cancer cells yet preserve healthy ones.

Of course, when you hear the word *"radiation,"* it is common to think of nuclear accidents and sci-fi movies where people glow in the dark. But in cancer therapy, radiation is one of the many tools doctors use to treat the disease. Radiotherapy uses high-energy beams, like X-rays, gamma rays, or Röntgen rays, to kill cancer cells. When undergoing radiotherapy to cure the tumor, it is essential to minimize the damage to the healthy tissue surrounding the affected area. For instance, when treating cancer of the left breast, it is essential to ensure that the heart receives minimal radiation to avoid any potential damage. Radiation therapy aims to eliminate cancer cells and does the least damage to healthy cells. For example, if you aim to treat cancer in the left breast, minimizing radiation exposure to the heart is essential. To accomplish this, radiotherapists and physicists, who are experts in radiotherapy, devise a treatment plan that considers the size, location, and type of cancer. They establish the parameters of the machines delivering the treatment to precisely target the affected area of the body. The treatment is typically administered over several weeks, and the type and stage of cancer determine the course of treatment. However, sometimes, it can also be shorter if the goal is "only" to lower cancer pain. Other types of radiation therapy also exist, such as radiosurgery, which is where patients get a very high dose of radiation on a small area, with almost the same curative effect as surgery.

Radiation therapy is often a necessary part of treatment, either as the only option or in combination with other therapies. A common combination is to use radiation therapy before surgery to shrink large tumors or afterward to kill any remaining tumor cells. For some time now, radiation therapy has even been given during surgery for some cancers, such as breast cancer. Radiation therapy aims to damage the cancer cells' DNA so they cannot grow or multiply. Another common combination is "adjuvant" (postoperative) radiotherapy, which is given after surgical removal of a breast tumor. This type of therapy helps to destroy any cancer cells that may have migrated close to the operated tumor, thereby curing the patient without having to remove the entire breast. Depending on the treatment goals, patients may need up to five radiation treatments per week. Each treatment session typically takes 5–10 s, but the entire process can take many hours. In some cases, the preparation time can be long. This may occur, for example, when a mask is used to protect sensitive organs of the head when treating brain metastasis with radiation therapy. Despite the time it takes, however, radiation therapy is often effective in treating cancer and is integral to most cancer treatment plans.

In patients who have received radiotherapy at an early age, regular monitoring of the treated area is recommended. Late tumors, known as "secondary to radiotherapy," are described as being slightly more frequent in these areas.

For example, a young woman who has had radiotherapy to the thorax as a child for lymphoma should have her breasts monitored earlier than usual. However, it's important to note that while radiotherapy sometimes exposes patients to potential risks in the future, it can also help cure a disease that can be fatal in the short term. As discussed earlier, the risks must always be weighed against the expected benefits, and the situation must be discussed in a multidisciplinary meeting before proposing this type of treatment.

Unfortunately, as radiation therapy can also damage healthy cells, side effects are a common concern. Such side effects can include fatigue, skin irritation, nausea, hair loss, and difficulty swallowing. The body cells most susceptible to side effects are the ones that divide quickly, such as skin, hair, and mucous membrane cells. Still, the good news is that most side effects only occur in the treated area. For example, when a patient is being treated for brain metastasis, the patient may lose their hair locally, or if a breast is being treated, the patient may have skin irritation that looks like sunburn in one place on the skin. Difficulty swallowing can also occur if a lesion next to the esophagus is treated. The process damages the skin cells, causing swelling, itching, redness, or dryness. For any skin irritation during your treatment, the radiologist can prescribe care products and ointments.

TIPS!

What should you discuss with your doctor before radiotherapy?

- What exactly will be irradiated and why?
- How long will the therapy last?
- What side effects should I expect during the therapy?
- How can I alleviate the side effects?
- What long-term side effects can result from the therapy?

While radiation therapy may have some unpleasant side effects, it is essential to remember that it is still an effective cancer treatment. Fortunately, most side effects are temporary and manageable and will disappear with time. If your doctor recommends radiation therapy, do not be afraid to ask questions and get additional information about what to expect. The potential benefits of radiotherapy are always weighed against the potential risks, and a multidisciplinary colloquium discusses the situation before proposing this type of treatment.

6.5   Systemic Treatments

Systemic treatments are designed to ***target cancer cells throughout the entire body.*** These treatments usually involve drugs that circulate through the blood-stream. The advantage of systemic therapy is that it can reach cancer cells that have metastasized or spread to other parts of the body. However, drugs face difficulty in reaching certain areas, such as the brain, which has a protective barrier system, and areas of the body with few blood vessels, which are essential because drugs are transported by blood vessels.

The downside of this approach, like others, is that healthy cells can also be damaged in the process. In addition, systemic treatments can have side effects that impact the entire body. For example, chemotherapy can cause fatigue, hair loss, and nausea. As a result, it is crucial to weigh the potential benefits and risks of systemic treatments before starting any therapy.

For a systemic treatment to be effective against cancer, it should specifically discern healthy cells from cancer cells and target only the last one. To achieve this type of targeting in the body, there is a field known as **drug delivery**, where formulation techniques are researched to transport drugs in a con-trolled and safe way to specific cells. Among the classical drug delivery formu-lations including tablets, powder, and injection, which result in multiple side effects due poor specific targetability, the implementation of nano-carrier sys-tems as drug delivery can overcome these drawbacks. Nano-carrier systems are particles with size ranges between 10 and 1000 nm; can have a synthetic or biological origin, can incorporate different types of drugs, from chemical to RNA; and keep the drug stable and safe during the transport to target cells. Various synthetic nano-drug delivery systems have been developed so far, including liposomes, micelles, dendrimers, and polymeric-based nanoparti-cles. Liposomes are extensively investigated and have achieved considerable clinical success, for example, liposomal for the treatment of breast, ovarian, and sarcoma have been used [2]. Due to their synthetic origin, liposomes show low biocompatibility; this means that our immune system recognizes and eliminates from your body [3]. This way the drug will not reach the can-cer cells and the treatment will not be effective. Biological nano-vehicles, including extracellular vesicles, outperform synthetic nano-vehicles because of their origin. Extracellular vesicles are released by every type of cell in our body and can exchange biological information between cells very efficiently. Their natural characteristic to transport different types of information can be used for drug delivery. Because extracellular vesicles are self-body products, they are less recognized by the immune system for degradation and can cross biological barriers and reach for example the brain, enhancing their potential for brain tumor treatment. To specifically treat cancer cells, the nano-carrier system needs to first recognize and deliver their content. Simply as sending a package to a friend, you have to specifically add an address on it, which helps postal services deliver it to the correct house without delivering it to the wrong location. Scientists have developed different kinds of strategies to achieve specific drug delivery, for example by incorporating antibodies, ligands, or proteins to the surface of nano-carriers that help the recognition and treatment of cancer cells [4, 5].

6.6   Chemotherapy

Chemotherapy is a drug treatment that prevents cancer cells from dividing correctly, thus causing their death. Therefore, the underlying principle of chemotherapy is that cancer cells are more sensitive to drugs than healthy cells because cancer cells divide more quickly. However, some *healthy cells divide quickly*, including those responsible for our immune system and the cells that keep our internal and external environments separate. These barriers must constantly repair themselves to function effectively, and their cells *divide rapidly* to achieve this. It is fascinating to see how chemotherapy drugs can specifically target cancer cells and protect healthy cells that don't divide rapidly, which is crucial for our overall health and well-being.

This means that when chemotherapy drugs are introduced into the body, they will preferentially target and *kill cancer cells while leaving healthy cells relatively unharmed*. Chemotherapy is generally effective at killing cancer cells. However, it can also damage fast-growing healthy cells, such as hair cells, blood cells, and mucous membrane cells. This can result in side effects like hair loss, skin dryness, reduced immunity, mouth ulcers, and digestive disorders. Although most of these side effects are temporary and resolve once the treatment is over, some patients may experience particular or late side effects that require close medical supervision. Chemotherapy is typically used in three situations:

1. *It can shrink or eliminate tumors* and in certain types of cancer even lead to a complete cure. With a partial response, the tumor can become small enough to make surgery possible or reduce the discomfort caused by troublesome metastases. Stable disease is another outcome where the cancer stops growing, which is a significant achievement. The ultimate goal is to

keep the tumor or metastases from growing or spreading so the patient can live a longer and healthier life.

2. **It can be used to kill invisible cancer cells.** Chemotherapy can kill cancer cells that are too small to be seen by the naked eye. This is crucial in situations where the cancer is aggressive and has a high risk of spreading. By administering chemotherapy before or after surgery, we can stop the cancer in its tracks and prevent it from coming back. We're committed to providing the best possible quality of life for our patients, and chemotherapy is an essential tool in our arsenal.

3. **It can be used to make radiotherapy more effective**, a process known as radiosensitizing chemotherapy. Before starting chemotherapy, it's essential to undergo a thorough medical checkup to determine whether there are any contraindications, especially when there is a risk of cardiac side effects. Close medical supervision during and after treatment is necessary to monitor for potential side effects, and treatment doses may need to be reduced or discontinued if concerning side effects appear. The ultimate goal of chemotherapy is to keep the tumor/metastases quiet for as long as possible, allowing the patient to live a progression-free life with the best possible quality of life.

When Do I Need Chemotherapy?

In cases where cancer cells are multiplying uncontrollably in the body, chemotherapy may be necessary to kill the cancer cells and prevent them from spreading. It can be given *before surgery* to reduce the size of large tumors, thus making the operation easier. If the drug is nonirritating, meaning it does not damage the veins, which can be painful, and the patient's veins are easily accessible, the infusion can be administered directly into an arm vein using a flexible tube called a catheter. However, if the drug is irritating or the veins are difficult for the nurse to locate, an alternative approach may be necessary. This could involve using a subcutaneous device such as a DAVI or cath port or an external device such as a PICC-line connected to a large chest vein. These devices are typically inserted during minor surgery.

Like radiotherapy, it can also be administered *after local therapy* to kill any remaining cancer cells and prevent cancer from returning. Furthermore, it is sometimes given as a preventive measure to minimize the risk of recurrence by killing any remaining tumor cells isolated in the body. Chemotherapy may also be given in palliative care. Even if cancer cannot be cured, it can stop the progression of cancer and allow the patients to live longer. In short, in most cases, it is administered *to cure or halt the disease.* However, chemotherapy can have side effects, such as fatigue, nausea, and hair loss. Before starting chemotherapy, it is critical to discuss its advantages and disadvantages with your doctor.

How Do I Receive Chemotherapy?

There are different forms of chemotherapy. If you are receiving *intravenous (IV) therapy,* you should know a few key points about the process. The drugs can be administered in different ways. First, they can be given directly as an infusion into the bloodstream, by drip through a tube into the arm vein or via a flexible tube in the neck. Second, if prolonged therapy is anticipated, a gentler method may be used by connecting the device to the large veins. The therapy is carried out at intervals according to a fixed schedule, usually on one or more days in a row. Finally, if you have any questions or concerns about the intravenous procedure, do not hesitate to ask your healthcare provider. Once you know more about it, receiving IV therapy can be relatively simple and straightforward.

Second, chemotherapy *tablets* have been on the market for several years now. However, not all proven cytotoxins are available in tablet form yet. In principle, even in tablet form, chemotherapy can be just as potent and effective as IV treatment. The significant advantage of chemotherapy tablets is that they can be administered at home without having to go to the hospital for each session. This not only saves time and money but also helps to reduce the risk of contracting an infection. Of course, self-administration does require some instruction and supervision from a healthcare professional beforehand. However, many find this option to be most effective and worth the effort.

With an increasing number of chemotherapy drugs becoming available in tablet form, it is hoped that even more people will benefit from this convenient and effective treatment option. Chemotherapy aims to destroy cancer cells while causing the least damage possible to healthy cells. However, chemotherapy drugs can damage healthy cells, leading to some side effects. Despite those potential side effects, chemotherapy is often an effective cancer treatment; it can help to shrink tumors and improve symptoms. In some cases, it may even cure cancer.

Side Effects of Chemotherapy

As discussed earlier, chemotherapy treatment can adversely affect rapidly dividing cells, especially those responsible for sealing the body's interior and exterior, such as skin, mucous membrane cells, marrow, and immune cells. The decision to stop chemotherapy and adjust the treatment dosage depends on the treatment's objective but primarily on the severity of the side effects. If the treatment aims to cure, we may encourage the patient to put up with troublesome side effects. However, if the aim is to improve the patient's quality of life, we may interrupt the treatment. When considering the side effects of chemotherapy, it's important to remember that there are different levels of gravity. Fortunately, many side effects are mild and do not require any change in the dosage. For example, grade 1 side effects such as redness on the soles of the feet following the intake of Capecitabine chemotherapy are not harmful to the patient. On the other hand, grade 3 side effects may require a temporary discontinuation of treatment until the effect diminishes or disappears. If the redness progresses and blisters form, preventing the patient from walking, a therapeutic pause is necessary. Be aware that the most frequent side effects of chemotherapy are those that impact the skin, mucous membranes, and immune system. Knowing this, the gradation of side effects can help you make reasoned decisions about your health.

Skin Cells

Chemotherapy can have a significant impact on skin cells, making them more fragile and prone to dryness and discomfort. Patients often experience dryness, itching, and flaking of the skin, which can be alleviated by daily application of moisturizing creams. In addition, the skin may become more irritated when exposed to natural irritants such as friction or the sun. To avoid further skin irritation, patients are advised to wear comfortable and loose-fitting clothing and avoid sun exposure when possible. Some chemotherapy treatments can also cause specific toxicities, such as *redness in the skin folds, palms, and soles of the feet*. In such cases, a pause in treatment and adjustment of the therapeutic dosage may be necessary. Chemotherapy can also affect the cells responsible for hair growth, sometimes resulting in *hair loss*. Depending on the treatment and the patient, hair loss can be gradual or occur rapidly within a week or two after the initiation of chemotherapy. Patients may start noticing more hair on their pillows, combs, or in the bathtub after washing their hair. The hair loss may affect all hair on the body, including

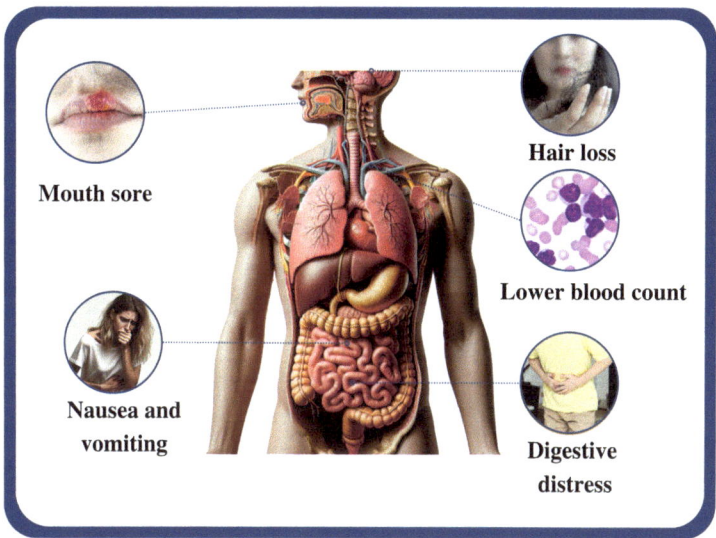

Fig. 6.5 Undesirable effects of chemotherapy. Chemotherapy also damages healthy, rapidly dividing body cells, thus causing various undesirable effects, including infection in different parts of the body

eyebrows, eyelashes, beard hair, and pubic hair. At this stage, many patients opt to shave their hair for practical and aesthetic reasons (Fig. 6.5). The good news is that hair usually starts to grow back within 1–4 weeks after the therapy, sometimes with changes in color, texture, and abundance. Some patients may experience white or gray hair growth, while others may find that their hair is curlier than before. Hair volume is often more abundant after treatment. The use of a cooling helmet during treatment can sometimes help preserve hair volume. The cold temperature constricts the diameter of the blood vessels, leading to the cells responsible for hair growth and partly preventing chemotherapy from reaching them. However, some chemotherapy treatments can cause systematic hair loss, even with a cooling helmet. In such cases, the oncologist will inform the patient of the expected outcome and help them prepare for it, for example, by acquiring a wig during this period. It's important to note that there is no correlation between side effects and the anticancer effects of a treatment. The choice of chemotherapy depends on the type of cancer, and the most effective complementary treatments for breast cancer almost always cause hair loss. On the other hand, chemotherapy given in combination with radiotherapy for rectal cancer does not cause hair loss. Therefore, it's essential to discuss the side effects of chemotherapy with the oncologist and understand the risks and benefits of each treatment option.

Mucous Membrane Cells, We Call It the Skin Inside the Body

Chemotherapy can harm the mucous membrane cells, resulting in their fragility. As a result, mouth ulcers and irritation can develop, leading to difficulties in swallowing and eating. The proliferation of yeasts, tiny fungi forming a white coating in the mouth, can also occur, leading to a loss of taste. The symptoms can be relieved with mouth rinses containing a little bicarbonate. The mucous membrane's fragility in the stomach can irritate, resulting in pain, burning, nausea, and loss of appetite, mainly when the belly produces acid for digestion. The symptoms can be minimized by suppressing acid production temporarily and taking anti-nausea medication. Irritation can also occur at the intestinal level, leading to discomfort during digestion, changes in stool consistency, and cramps. The treatment and dietary advice offered to the patient depends on the adverse effects and the degree of discomfort.

Bone Marrow and Immune Cells

The cells circulating in our blood are born and reproduced in the bone marrow, which, interestingly, is the fatty substance inside bones that some of us have eaten.

These cells divide rapidly to ensure that there are always enough red blood cells, which are responsible for carrying oxygen throughout our body and can be compared to little oxygen delivery trucks; white blood cells, acting as soldiers who constantly fight off any infections; and platelets, a type of cell without a nucleus that help to repair wounds. A lack of red blood cells may result in anemia, which can cause specific vital organs to suffer from a lack of oxygen. Similarly, if there aren't enough white blood cells, even a simple microbe could become a real danger, and if there aren't enough platelets, even a tiny wound could bleed profusely. It is important to note that like red blood cells, white blood cells, and platelets, the ones in the bone marrow are also susceptible to chemotherapeutic drugs, which may temporarily interrupt the production of new blood cells during treatment.

This is why it's essential to keep an eye on the quantity of blood cells by taking regular blood samples during chemotherapy. This helps to adjust the doses or the interval between chemotherapies if necessary. The administration of chemotherapy usually ensures that the level of these blood cells is sufficient. The pause between each treatment, known as a "cycle," typically lasts between a few days and a few weeks, allowing the marrow to recover from the damage caused by each treatment. Generally, a treatment cycle lasts for 3–4 weeks. Clinical studies have established the ideal dose and pause between each treatment to obtain the best effect with the fewest possible side effects. However, it's essential to adapt the treatment to each patient. No point in continuing chemotherapy if it isn't working, and it's advisable to reduce the dose of a medicine that's too toxic for a particular patient. Chemotherapy temporarily weakens the body's defenses by reducing the immune cells, as shown in (Fig. 6.6). Nonetheless, recent medical advances have made it easier to manage the side effects of chemotherapy. New drugs are regularly developed to help improve patients' tolerance to chemotherapy. Most often, chemotherapy can be administered on an outpatient basis, meaning that there's no need for an overnight stay in the hospital. However, in some cases, it may be necessary to stay in the hospital for a few days to avoid certain risks of complications or to enable the treatment to be given continuously over several days. This is done, for example, to ensure that patients are properly hydrated and to prevent the risk of renal toxicity of certain products. It's important to remember

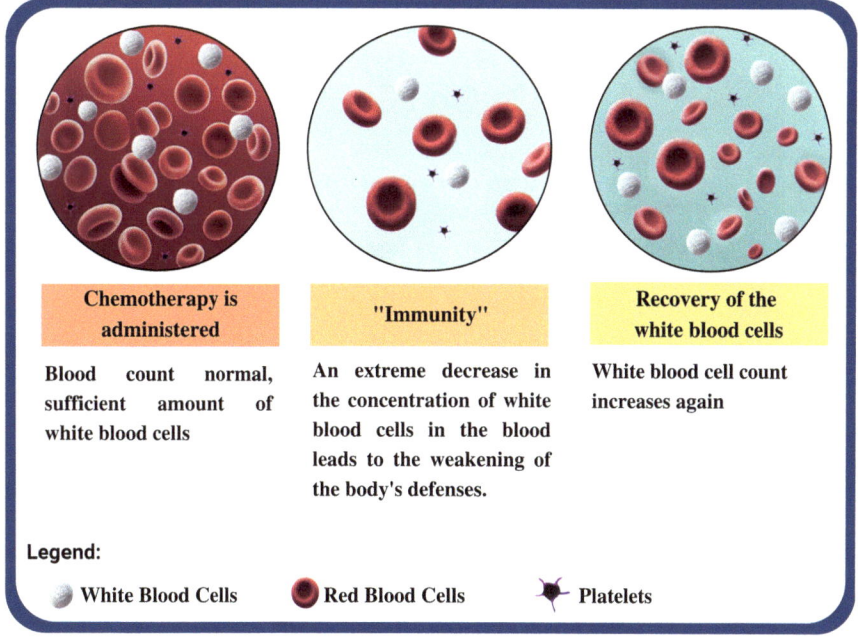

Fig. 6.6 Effect of chemotherapy on the white blood cells. Chemotherapy temporarily weakens the body's defenses since it reduces the immune cells

that chemotherapy is a critical part of the treatment for many types of cancer. Most people tolerate chemotherapy well and experience only occasional unpleasant side effects. These side effects may temporarily impair patients' quality of life, but they are usually minor and short-term. However, high-dose chemotherapies administered to prepare certain patients for transplantation or to treat certain leukemias, for example, can place patients in highly vulnerable situations.

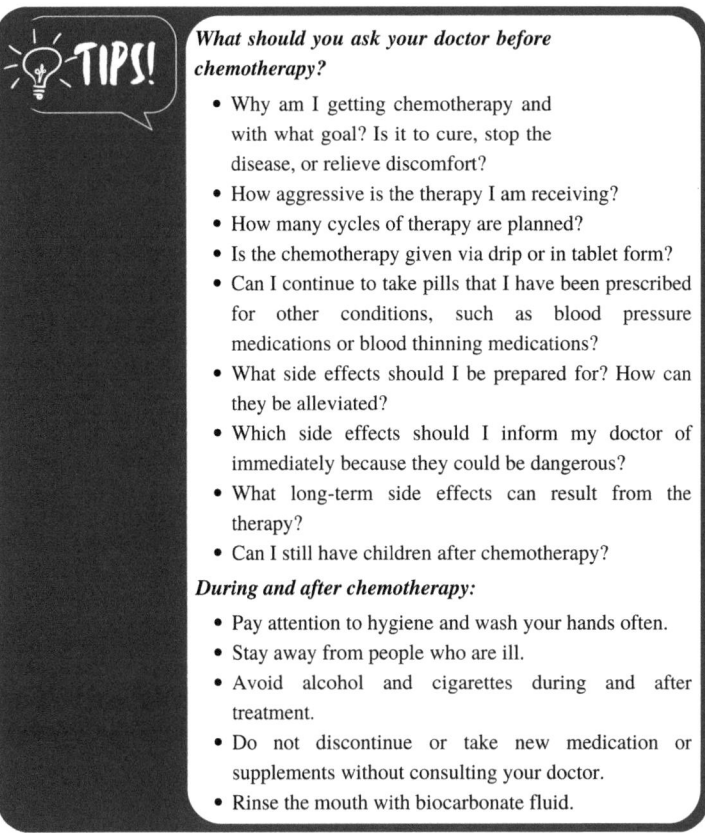

TIPS!

What should you ask your doctor before chemotherapy?

- Why am I getting chemotherapy and with what goal? Is it to cure, stop the disease, or relieve discomfort?
- How aggressive is the therapy I am receiving?
- How many cycles of therapy are planned?
- Is the chemotherapy given via drip or in tablet form?
- Can I continue to take pills that I have been prescribed for other conditions, such as blood pressure medications or blood thinning medications?
- What side effects should I be prepared for? How can they be alleviated?
- Which side effects should I inform my doctor of immediately because they could be dangerous?
- What long-term side effects can result from the therapy?
- Can I still have children after chemotherapy?

During and after chemotherapy:

- Pay attention to hygiene and wash your hands often.
- Stay away from people who are ill.
- Avoid alcohol and cigarettes during and after treatment.
- Do not discontinue or take new medication or supplements without consulting your doctor.
- Rinse the mouth with biocarbonate fluid.

Because of recent medical advances, treating chemotherapy side effects is easier and more manageable. In most cases, chemotherapy can be done as an outpatient. However, an increased risk of complications, or the way the treatment is given, might require hospitalization. Regardless of the severity of the side effects, it is important to remember that chemotherapy is an essential part of treatment for many types of cancer. What is more, with proper care and treatment, most people tolerate chemotherapy and experience only minor side effects. For elderly patients, there are scores that can be used to calculate the risk of complications, making it easier to weigh up the pros and cons of introducing treatment.

A promising way to attenuate the side effects of chemotherapy is by using antibody drug conjugate (ADC), known as "biological missiles." [6] One of the key properties of antibodies is their ability to recognize specific antigens allowing them to precisely target cancer cells, much like a missile identifies its target. By linking an antibody to a potent drug, such as a chemotherapeutic agent, an effective "killing" of cancer cells can be achieved. Currently, there are 11 ADCs approved by the Food and Drug Administration (FDA), and over 100 are undergoing various stages of clinical research. This targeted therapy approach has proven to deliver more successful antitumor effects compared to conventional chemotherapy, especially in the treatment of lymphoma, breast, gastric, lung, urothelial, cervical, and ovarian cancers [7]. Regarding the side effects of chemotherapy previously explained, ADCs have a significant advantage because they target cancer cells exclusively without damaging healthy cells in the body.

Why Chemotherapy Can Fail

Why chemotherapy sometimes doesn't work: cancer chemotherapy can fail for various reasons. One is that cancer cells may have developed resistance to chemotherapy [8]. They may not let the drug "in," favor its expulsion, or render it ineffective when it crosses their surface. They may also not receive chemotherapy at all because of their location (the brain, e.g., is a complex organ to reach with chemotherapy) or because they are far from the blood vessels (which is essential to deliver the treatment). According to research, tumors are often heterogeneous, meaning they are made up of different cells, and it only takes a few resistant cells in cancer for chemotherapy to fail. Even if all the neighboring cells have been destroyed, the immune cells will continue to develop. When the majority of cells are sensitive to chemotherapy, but there are a few resistant cells, chemotherapy first leads to the destruction of susceptible cells, which reduces tumor size, and then to an increase in tumor size as resistant cancer cells develop, taking the place of the destroyed cells. The tumor then "grows back" after initially shrinking. This explains why, on a CT scan, chemotherapy may appear to work initially, only for the cancer to increase again later. If all the cells are resistant, there will be no destruction, and the tumor will continue to grow despite the chemotherapy [9].

6.7   Hormone Therapy

Hormones are natural substances that are produced by glands and circulate in the blood, transmitting information throughout the body. The information that hormones transmit can differ according to the hormones and the organs that will receive the information. This means that cells will only receive the information if they have a receptor to obtain it, just like we can only hear information if we have ears. The intricate and fascinating complexity of cell reactions to the same hormone is underscored by the fact that they can vary depending on the organ they belong to. Similarly, the same order (e.g., "do your job!"), although received by the same receptors (ears), will be carried out differently by a baker than a gardener, illustrating the captivating nature of this process.

Receptors, often referred to as "ears," are pivotal in cell reactions to hormones. Consider tumor cells with receptors on their surface, which can be sensitive to messages transmitted by hormones. In the context of cancer, it's crucial to prevent the production of "message hormones" that could potentially "block the ears" or receptors. This could lead to uncontrolled cell growth, underscoring the importance of understanding this process. If the message is to "grow," it's easy to see why it's essential to prevent the production of the hormone messages or to block the receptors. Hormone therapy, also known as hormone modulation, is a cancer treatment that blocks the action or manufacture of hormones that tumors need to grow. By doing so, hormone therapies can help prevent the spread of cancer and improve the patient's chance of survival.

Hormone therapy is a cancer treatment that either ***blocks the hormones*** that tumors need to grow or ***prevents the production*** of these hormones in the body. Hormone therapy is also known as hormone modulation therapy or hormone treatment. Some tumors depend on hormones so they can grow in the body. For example, breast cancers can be fueled by estrogen, while prostate cancer is driven by testosterone. The right hormone therapies can lower the levels of those hormones or block their effects on tumor cells. Hormone therapy is frequently combined with other treatments, such as surgery or chemotherapy. In some cases, it may be used as the only treatment for cancer.

In the context of breast cancer, it will be analyzed by examining the biopsy or surgical specimen to determine whether it is of the "luminal" type, meaning the cancer cells are sensitive to estrogen hormones. The pathologist will examine the specimen under a microscope to see if hormone receptors have developed on the cancer cells. These hormone receptors act like locks, into

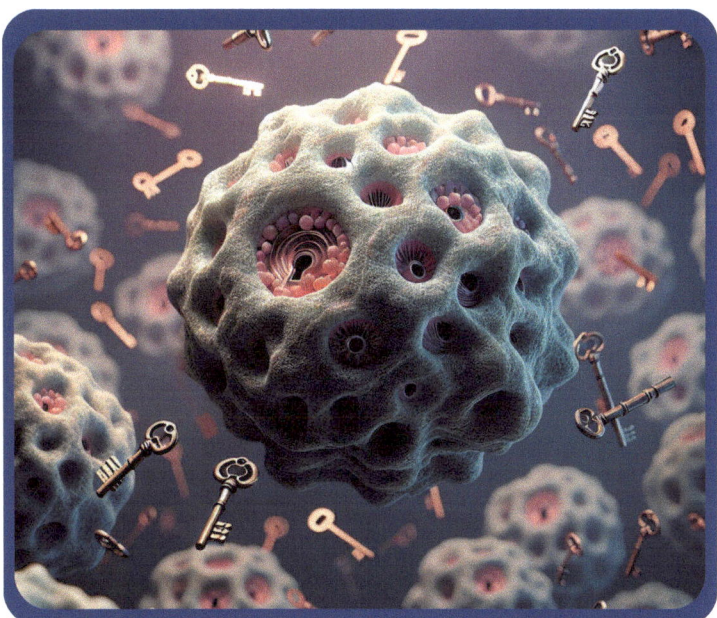

Fig. 6.7 The image shows an illustration of a cell with numerous keyholes on its surface surrounded by floating keys. This visual metaphor hints at the concept of the lock-and-key principle, which is often used to describe how specific molecules, such as hormones and receptors, interact with each other in biological systems. The keys represent the hormone, which can interact with the cell by fitting into the keyholes that symbolize receptors. This illustrates the specificity and selectivity of hormones, which are crucial for processes such as cellular communication and metabolic pathways

which estrogen can be inserted like a key, which can trigger cell proliferation (Fig. 6.7). If these hormone receptors are present, antihormone treatment can be used, which involves administering two types of antihormonal drugs. One type of drug blocks the receptor locks and prevents essential hormones from entering, while the other type prevents the production of key hormones. However, if the hormone receptors are not present, there is no point in giving hormone therapy as the tumor cell divides independently of hormones. It is also possible for luminal tumors to become resistant to hormone modulation, which requires modifying the treatment plan.

By understanding the importance of hormone receptors and the effectiveness of antihormonal treatment, patients can take an active role in their fight against breast cancer. Together with their healthcare providers, they can make informed decisions about the best course of treatment. In prostate cancer, the principle is the same. However, it is not necessary to check for the presence of testosterone receptors, as almost all prostate cancer cells express these receptors on their surface. Hormone therapy is effective for the vast majority of luminal breast cancers and prostate cancers, although the effectiveness may be temporary. In both these situations, the most frequent side effects are frequently similar to those experienced when hormones drop, as in menopause for women or andropause in men - a less well-known but equivalent condition caused by decreasing testosterone levels.

While there is no way to eliminate the side effects of hormone therapy, some things can be done to minimize them. Hot flashes and night sweats can be reduced by avoiding spicy food and wearing loose-fitting clothing. Acupuncture and integrative medicine can often help alleviate some symptoms. Depressive phases and mood swings can be alleviated by talking to a trusted friend or counselor or even by taking medication if necessary. It is also essential to inform your doctor about any bothersome symptoms so they can be addressed appropriately. A patient's quality of life can be improved by changing hormone therapies, as not all drugs have the same side effects.

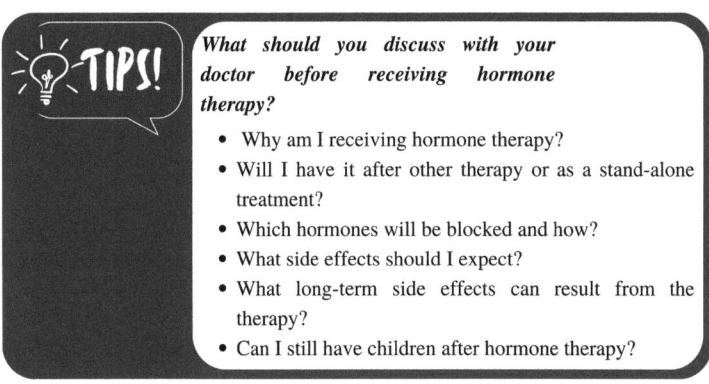

What should you discuss with your doctor before receiving hormone therapy?

- Why am I receiving hormone therapy?
- Will I have it after other therapy or as a stand-alone treatment?
- Which hormones will be blocked and how?
- What side effects should I expect?
- What long-term side effects can result from the therapy?
- Can I still have children after hormone therapy?

6.8   Immunotherapy

The Immune System and Cancer

Our immune system is unique. It is composed of different cells, proteins, tissues, and organs working together to protect us. White blood cells are one of the most important parts of the immune system. They help fight off infection by bacteria, viruses, and fungi (Fig. 6.8).

The immune system works by recognizing the body's cells and attacking and destroying anything foreign it does not recognize. It does it as a precaution. The main goal is to maintain a perfect balance (homeostasis) between all

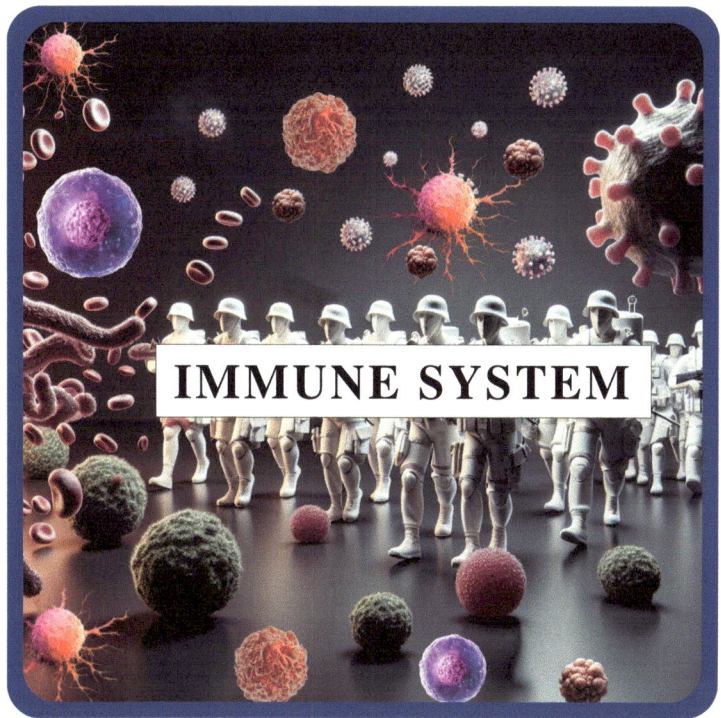

Fig. 6.8 The image creatively depicts the immune system as a battalion of soldiers, symbolizing its role in defending the body against pathogens. Immune cells: The soldiers represent various immune cells, such as T cells and macrophages, that protect the body.Pathogens: The soldiers are surrounded by viruses, bacteria, and other pathogens that the immune system fights and neutralizes.This metaphor illustrates the function of the immune system as the body's defense mechanism, constantly monitoring and responding to potential threats to maintain health and prevent infection

elements of the body. The immune system tries very hard to maintain this balance and return to it when it is lost. A well-functioning immune system will also attack and destroy cancer cells before they have a chance to increase.

Unfortunately, cancer cells can seem very similar to healthy cells. They have "masks of normalcy" on their surfaces, enabling them to hide and evade the immune system. This makes it difficult for the immune system to recognize them as foreign or dangerous. In the presence of tumors, immune cells can appear "calm" or "exhausted," as if they are restrained from attacking, almost like they are influenced by inhibitory messages from the tumor. In a way, it's as if the tumor "paralyzes" the immune cells. That makes it hard for the immune system to recognize them as foreign invaders. If our defenses are tricked, cancer can spread undetected. In addition, some cancer cells consistently mutate, making it even harder for the immune system to keep up [10]. But despite these challenges, our immune system normally does an excellent job of keeping us healthy. Cancer cells are clever mechanisms. They know how to evade the body's immune system in order to continue its rapid and unchecked growth. However, since cancer is a constant danger and threat, our immune system is also constantly learning how to cope with it. Most of the time, the protection is successful. But we must worry about the remaining 0.01%, where it is caught off guard, and a single cancer cell becomes life-threatening.

Our immune system is like an army, with different kinds of soldiers and leaders. Our different immune cells work together to defend us from cancer. When cancer cells grow uncontrollably, the immune cells usually get curious and go to the scene to see what is happening. Upon recognizing cancer cells, natural killer cells such as white blood cells immediately get to work by killing these cancer cells. Then, the defenders send a signal to activate the danger mode of the dendritic cells, which will collect samples of the dead cancer cells. They will also activate the T cells in the lymph nodes to block new blood vessel growth and starve the cancer cells, killing them by the hundreds. Lastly, the macrophages eliminate all the cancer cell corpses. However, our immune cells must do their jobs accurately and efficiently because even if one cancer cell survives, it will find ways to fight back. This single-survivor cancer cell would acquire more mutations, grow and multiply rapidly, and become more uncontrollable. They will then develop into even better at hiding from the immune system.

What's more, cancer cells can find ways to paralyze our immune system. They do this by deactivating killer T cells and natural killer cells before they can start their attacks on the cancer cells. They thus stop them from killing. Now, the tumor can produce thousands of clones that are all resistant to the cancer cells and are not visibly defective. Cancer cells are also silent, which means they shut down the immune system using faulty signals. When cancer has overwhelmed the immune system, the tumor will grow much larger, this time becoming a big problem.

It's interesting to note that the ability of tumor cells to disable the immune system plays a crucial role in the growth and progression of cancer. Tumors that possess a high ability to paralyze the immune system can grow unchecked. In contrast, those that lack this ability or have it blocked can be rapidly destroyed and never cause cancer. Now, the big question is, how can we "unblock" the immune system that has been rendered ineffective by tumor cells? Well, recently, treatments have been developed that aim to reactivate a "blocked" immune system. These treatments are collectively known as immunotherapy. ***Immunotherapy*** uses the body's defenses to outsmart cancer cells and stop them from growing and spreading. The goal of immunotherapy is to get the immune system to recognize and attack cancer cells while leaving healthy cells alone [10]. Immunotherapy comes in two forms: active and passive. ***Active immunotherapy*** involves stimulating the immune system to produce more cancer-killing cells. ***Passive immunotherapy*** involves injecting antibodies (a protein component of the immune system) into the body. The antibodies act like arrows, targeting and binding specifically to the surface of tumor cells. It's as if they are magnetized, and the cells have metal locks that only the antibodies can unlock (Fig. 6.9). If the cell doesn't have the specific surface element that matches the antibody, it won't bind to it. By blocking the cell's metal lock, the antibody can prevent hormones, growth factors, and other molecules from activating mechanisms that promote tumor growth. Additionally, the antibody can mark the cancer cells and signal white blood cells to destroy them. The chapter on targeted therapies will provide additional information on this topic, including "passive" immunotherapy.

Only a few immunotherapies are currently available, but many more are being researched in clinical trials.

Passive immunotherapies are created using identical antibodies, also known as "monoclonal" antibodies in medical terminology, abbreviated as MAB. These identical antibodies have been designed to target specific proteins, making them highly effective in treating various types of cancers.

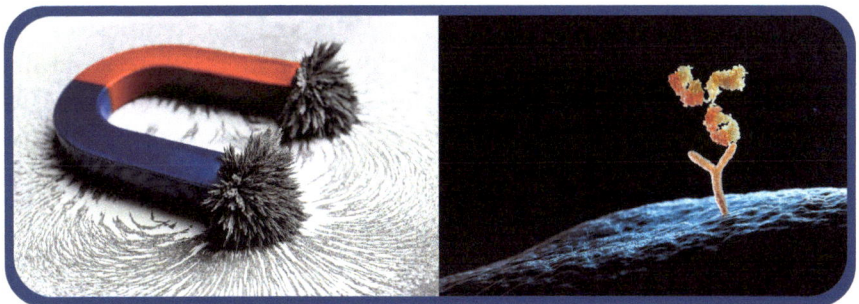

Fig. 6.9 On the left is a red and blue horseshoe magnet attracting iron filings. The filings are arranged in patterns that display the magnetic field lines around the magnet. On the right is an illustration of an antibody. The antibody is depicted as a Y-shaped structure on a cellular surface, demonstrating its role in the immune system. Antibodies are proteins that recognize specific antigens and bind to them, helping to neutralize pathogens like bacteria and viruses. The image draws an analogy between the binding force of an antigen and antibody and that of a magnet attracting iron filings. Just as magnets have a specificity for iron, an antibody has a specific binding affinity for a particular antigen (protein structure) due to its surface structure (epitope)

RituxiMAB, TrastuzuMAB, and PertuzuMAB are just a few examples of the many monoclonal antibodies that have been developed and are being used to fight cancer. With their proven track record of success, MABs are a powerful tool in the fight against cancer, giving hope to many patients and their families.

Immunotherapies differ from traditional cancer treatments such as surgery, radiation, and chemotherapy. Those treatments work by killing all fast-growing cells, including cancer cells. However, as we mentioned in previous chapters, they can also damage healthy cells, which can cause side effects like hair loss, nausea, and fatigue. Immunotherapy activates the immune system to fight cancer cells. However, it's important to note that immunotherapy may not be effective for all cancer types, and some patients may experience distressing side effects. The decision to use immunotherapy as a treatment option is made on a case-by-case basis, considering the type of cancer and the patient's specific situation. Receptors found on the surface of cancer cells can also help determine whether immunotherapy is likely to be effective. So far, immunotherapy has helped many cancer patients live longer and feel better. Immunotherapy holds great promise as a treatment for cancer. This is why scientists are working hard to develop new and improved immunotherapies.

A research team in Heidelberg has presented a ground-breaking development in this field. They have developed an algorithm called "predicTCR" that uses artificial intelligence to identify T-cell receptors that are likely to react to tumor characteristics. This approach could significantly simplify and accelerate the identification of personalized tumor-reactive T-cell receptors. The team isolated tumor-infiltrating lymphocytes from a melanoma patient's brain metastasis, sequenced them individually and tested their receptors. Based on this data, they were able to train an AI model that can identify tumor-reactive T cells with 90% accuracy. The development of "predicTCR" is a major step forward in cancer treatment. With this technology, personalized cellular immunotherapies can be produced more efficiently, improving the chances for cancer patients. The combination of immunotherapy and AI offers new hope for cancer treatment. By better understanding and harnessing the immune system, we can develop more effective therapies that improve the lives of cancer patients [11].

Immune Checkpoint Therapy

The immune system is a fascinating thing. It comprises various cells working together to protect the body from foreign invaders. One type of cell, known as a *T cell,* is responsible for destroying pathogens. However, T cells must pass through a series of ***checkpoints*** before they can do their job. These checkpoints are designed to prevent the T cells from attacking the body's healthy cells. The organs most frequently affected are the digestive tract (patients develop diarrhea), the lungs (patients develop pneumonia), and the skin (patients develop rashes). However, if these checkpoints are stimulated by other cells, they send a signal to the T cells, halting any form of attack (Fig. 6.10). This process is crucial to preventing immunological attacks. Without it, our bodies would constantly be attacked by our own immune systems!

Immune checkpoint inhibitor drugs are a new and revolutionary cancer treatment. These drugs prevent cancer cells from suppressing the immune response. They do this by controlling the blockade of checkpoints. This enables the T cells to recognize and destroy the cancer cells and thus to function normally.

Immune checkpoint therapies are frequently administered in an infusion solution at intervals of 2–4 weeks, depending on the drug used and the cancer type. Immune checkpoint therapies can have long-term effects as T cells can recognize and fight cancer cells for months or even years after receiving immune checkpoint drugs. A combination of multiple checkpoint inhibitors may also be used in treatment. For certain cancer types, immune checkpoint therapy can be combined with antibody therapy (Fig. 6.11), radiotherapy or chemotherapy [12]. This new and innovative cancer treatment is changing lives and giving hope to everyone worldwide.

Side Effects of Immunotherapy

When the body overreacts to any form of stimulus, more harm than good can be done. When it comes to the immune system, this is particularly true. While the immune system protects the body from disease and infection, an excessive immune reaction can damage healthy organs and tissues. Immune cells can sometimes mistake healthy cells for foreign invaders and attack them accidentally. If this happens, doctors may need to lower the dose of immunotherapy or stop it altogether. They may also need to give treatments that lower reactivity from overly aggressive immune cells, such as cortisone. Doing so can help prevent further damage and ensure the body remains healthy and balanced (homeostasis).

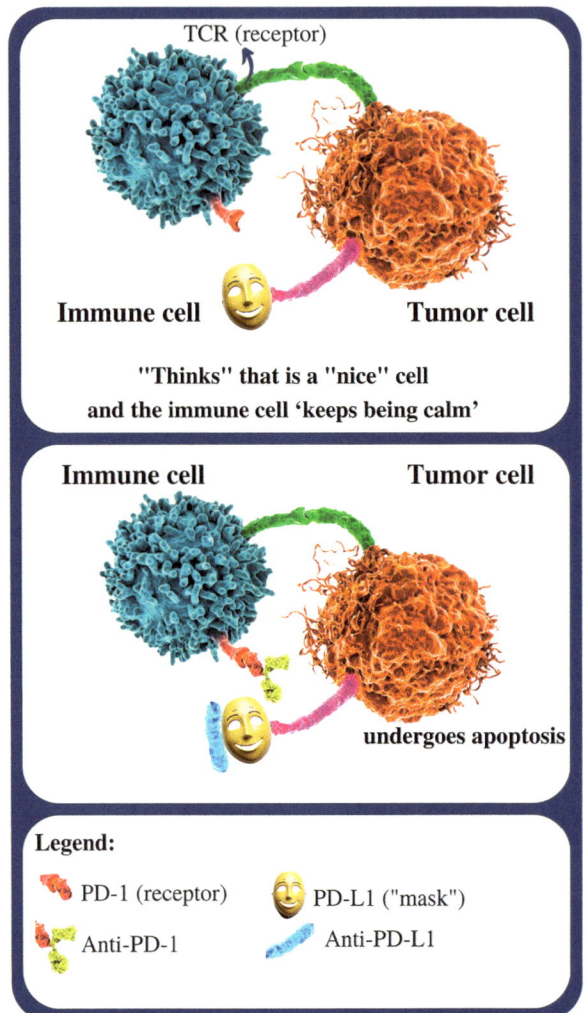

TCR (receptor)

Immune cell   Tumor cell

"Thinks" that is a "nice" cell
and the immune cell 'keeps being calm'

Immune cell   Tumor cell

undergoes apoptosis

Legend:

PD-1 (receptor)   PD-L1 ("mask")

Anti-PD-1   Anti-PD-L1

Fig. 6.10 Cancer cells evade immune system detection. Some cancer cells do not exhibit definite characteristics known as antigens or undergo mutation very often. Thus, they can deceive the immune system or be invisible to the receptors of immune cells and, thus, eventually escape

Targeted Therapies

Targeted therapies involve using drugs or other substances to block cancer cell development. This is done by interfering with specifically targeted proteins that are involved in cancer's transition, progression, and spread. There are many targeted therapies, and more are being developed. Immunotherapy is

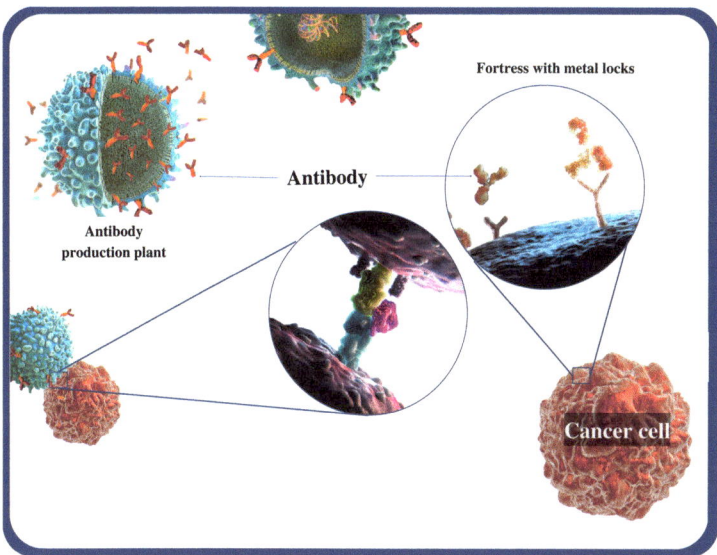

Fig. 6.11 Antibodies similar to those produced naturally by our white blood cells can be manufactured industrially. These antibodies are like little y-shaped arrows. Synthetic antibodies bind selectively to the receptors of cancer cells, just as key binds to the corresponding padlock. In this way, they can block the receptor (and thus render it ineffective), deliver treatments directly to the tumor cell, or activate the cell death process, for example

one type of targeted therapy that uses drugs or substances to stimulate or suppress the immune system's response to cancer. Another type of targeted therapy uses small drug molecules that enter the cell through the cell membrane and interfere with specific molecules involved in how cancer grows or spreads. Still, others deliver radiation directly to cancer cells.

Targeted therapy is also called precision medicine because it is based on identifying specific changes in cancer cells that make them more likely to respond to treatment. Targeted therapies target the altered proteins, enzymes, and receptors in cancer cells that cause them to grow, divide, and spread [13] (Fig. 6.12). Some targeted therapies focus on molecules in a cell's ability to divide and grow. Others stop cancer cells from making new blood vessels, which they need to succeed.

Unfortunately, not all cancers have specific mutations that can be targeted. When a mutation can be targeted, not everyone with that change will respond to a targeted treatment. Furthermore, because targeted therapies block a particular protein, cancers often find a way to keep growing by changing (mutating) so they no longer have the same protein targets. Even when these drugs work well at first, most cancers eventually find a way to grow and spread

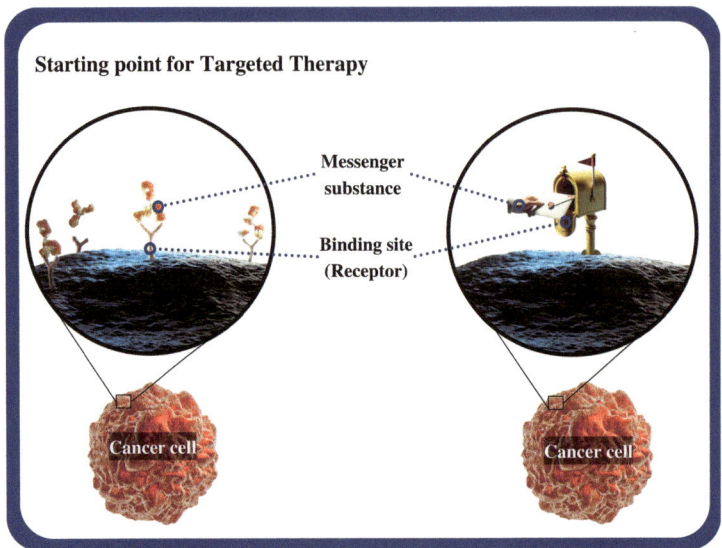

Fig. 6.12 Targeted therapies use drugs to target specific tumor characteristics. These specific characteristics include receptors on the cell surface, messenger substances that transmit signals to cancer cells, and signaling pathways

despite them. That is why it is crucial to develop new targeted therapies and combine them with other treatments. When used alone, targeted therapy does not typically cure cancer, even if that may happen in some cases. Instead, it can control cancer for a long time. Targeted therapy is frequently used with other treatments, such as surgery, radiation therapy, and chemotherapy, to achieve the best results possible. When breast cancer biopsies "show" expression of the HER-2 target, there's a good chance that treatment with anti-HER-2 antibodies will work. In lymphoma, if the abnormal cells express the "CD20" receptor, therapy with "anti-CD20" antibodies will work. The length and choice of treatment depend on the type of cancer, the drug given, and findings from sequencing methods such as NGS [13].

A recent study indicates that preventive treatment with the aromatase inhibitor [14] anastrozole may be particularly effective for women at higher risk of breast cancer, especially those with elevated estrogen levels. These findings were presented at the San Antonio Breast Cancer Symposium. Anastrozole [15] has already been approved for breast cancer prevention in the UK. The study suggests that estrogen levels could serve as a straightforward criterion for selecting patients for this preventive treatment.

One significant barrier to treating cancer successfully is the lack of targeted therapy options. For targeted therapy to be effective, the cancer cells need protein structures that can be targeted. Doctors must first test for their presence, which means that tissue must be obtained through biopsy, surgery, or blood.

However, targeted therapy has shown great promise recently. It may eventually provide a more effective way to treat cancer. Some people might think that drugs are only used to treat disease once it has progressed to an advanced stage. While this is often the case, many drugs can be used *in the early stages* of the disease to stop its progression and relieve symptoms. These drugs are available in various forms, such as *infusions, tablets, and injections*. They can be used alone or together with other therapies [13]. Targeted therapies are a newer type of drug that is still being researched. The examples below are treatments still under investigation and given only in research settings. However, they show promise as potential treatments for various diseases.

Signal transduction inhibitors are cancer treatments that block several important signaling cascades in the cancer cell. These drugs interfere with the way cells communicate with each other, and they prevent cancer cells from receiving specific signals that promote their growth. As a result, signal transduction inhibitors can disturb the growth of tumors. These drugs are the therapy of choice in treating certain leukemia, lung, breast, colon, kidney, skin, thyroid, and liver cancers. Fatigue is the most common side effect of signal transduction inhibitors. Other side effects include bleeding, infection, myelosuppression, and gastrointestinal issues. These drugs can be used alone or together with other anticancer agents (Fig. 6.13).

One promising treatment is the use of *growth inhibitors*. Growth inhibitors occupy the cell surface receptors that are responsible for receiving signals sent by growth factors. By preventing the growth factors from sending signals and binding to receptors, cancer cells can no longer receive the instruction to increase in size [8]. This can significantly reduce a tumor's size. Growth inhibitors may also promote cell death, further reducing the number of cancer cells overall. While more research is needed, growth inhibitors show potential as a new treatment option (Fig. 6.14).

Angiogenesis inhibitors are drugs that target new blood vessel formation. This inhibition is important because tumors cannot grow without a constant supply of blood, oxygen, and nutrients. Inhibiting angiogenesis means damaging cancer's supply lines. This treatment can effectively prevent the growth of cancerous tumors [13]. Many types of angiogenesis inhibitors are being developed today, and each one works slightly differently. Some inhibit the production of vascular endothelial growth factor (VEGF), and others block the protein activity that promotes blood vessel formation. No matter what

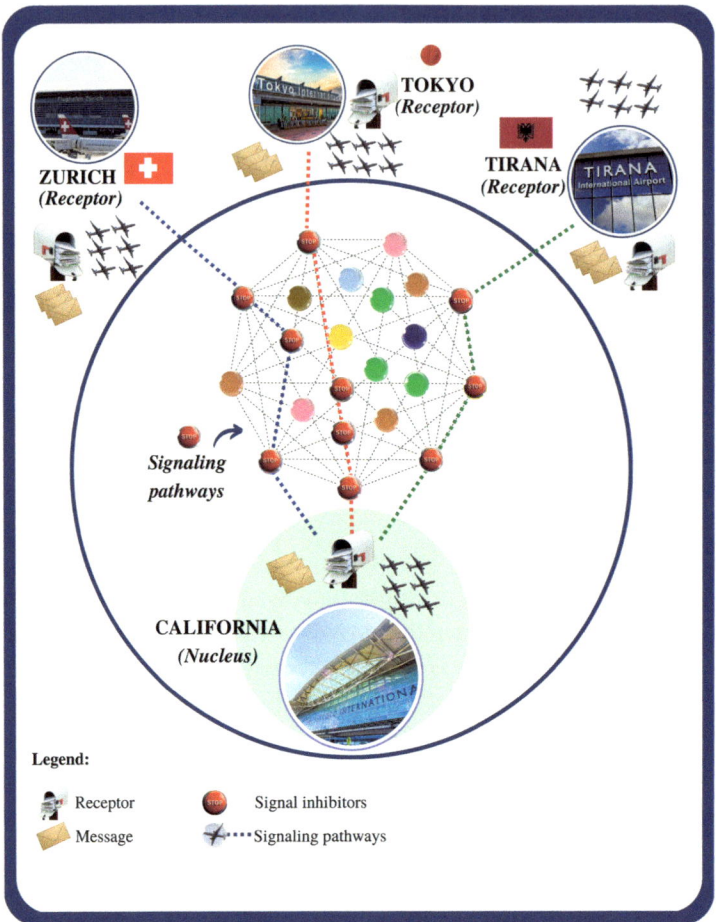

Fig. 6.13 Signal transduction inhibitors. Bad weather can stop planes from flying due to safety reasons. Similarly, signal transduction inhibitors block signaling pathways to stop tumor growth

their function is, all angiogenesis inhibitors share the same goal: to stop cancer in its tracks.

Monoclonal antibodies are unique drugs that are used to fight cancer. Monoclonal antibodies recognize and bind to specific antigens on the surface of tumor cells. Once the monoclonal antibody is "docked" on the cell, the immune system can recognize cancer cells and fight them. Some chemotherapy molecules can be fixed to those antibodies and target cancer cells specifically. These drugs are indeed a wonder of modern science!

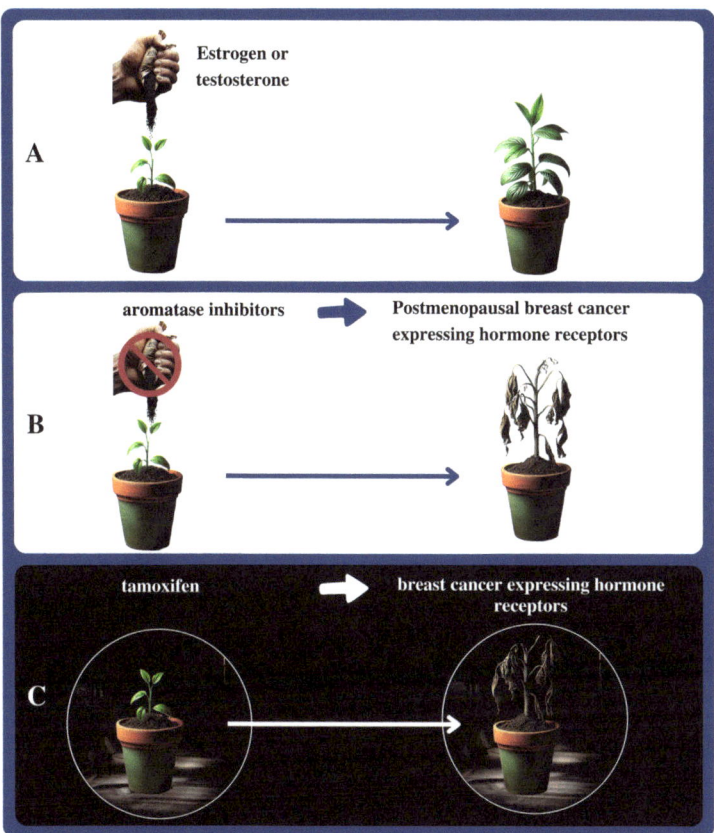

Fig. 6.14 Growth inhibitors. (**a**) Growth factors are like fertilizers that make the seed grow. If fertilizers aren't added (**b**), or if fertilizers can't reach the seed (**c**), the seed can't grow. Similarly, if the growth factors can't reach the cancer cell, it can't grow

6.9 New Technologies (Hope)

mRNA Therapy

The use of mRNA technology, which involves using messenger RNA to produce proteins, shows great promise in cancer treatment [16]. This innovative approach is especially effective when combined with immunotherapy, such as creating a personalized vaccine for each patient based on the specific antigens found in their tumor tissue. Initial results indicate that this personalized therapy can extend the period of time before cancer recurs compared to traditional methods, offering hope for further advancements in cancer treatment.

Another promising approach involves injecting mRNAs directly into tumors to produce cytokines [17], which could activate the body's immune system against palpable tumors like malignant melanomas and head and neck tumors. Combining this injection with checkpoint inhibitors, a type of drug that helps the immune system recognize and attack cancer cells, could further boost the immune response.

In addition to oncology, RNA-based substances also hold promise for treating wounds in diabetes patients and enhancing heart function in heart failure [10]. This broad potential prompts curiosity about the diverse applications of these technologies, inviting further exploration and research.

Research shows that micro-RNAs, small RNA molecules that regulate gene expression in cancer cells, can provide insights into personalized drug treatment options [16]. Studying these micro-RNAs can help identify genes that promote cell growth, leading to more targeted therapies and improved cancer treatment.

Research in mRNA technology and microRNA analysis presents promising avenues for the development of cancer therapy, with potential applications in various medical fields. An example of this is the mRNA vaccine autogene Cevumeran, which is showing promising results in the fight against pancreatic cancer. These findings could potentially herald a new era in the treatment of this disease [18], capturing the attention of the medical community.

An Example of the Pancreas

The pancreas plays a crucial role in digestion by producing enzymes that break down proteins, fats, and carbohydrates in the duodenum. However, in hereditary pancreatitis, the enzyme trypsin becomes active within the pancreas itself, leading to self-digestion of the tissue. This ongoing damage triggers inflammation and healing processes that can result in malignant pancreatic cancer in 40% of those affected [19]. Common causes include excessive alcohol consumption, gallstones, or elevated blood lipid levels.

Imagine having a key that reveals how a specific tumor will respond to treatment. Researchers have developed such a key by growing small "organoids" from tumor cells in the lab. These organoids mimic the behavior of the actual tumor within the body (Fig. 6.15). A new model, akin to a computer program, uses information from these organoids to predict the effectiveness of cancer therapies in patients with pancreatic cancer.

This model has been tested on organoids from various patients and has shown promising results. For patients who haven't yet received treatment, the

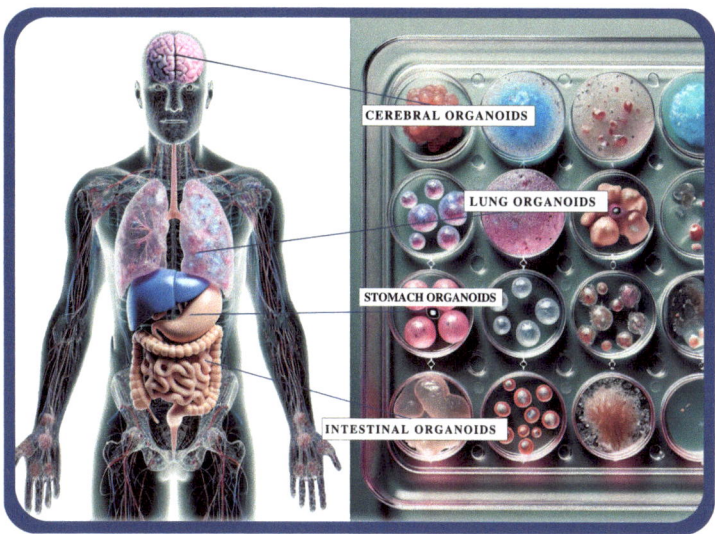

Fig. 6.15 This image illustrates the concept of organoids and their relationship to human organs. On the left, a human body is shown with the brain, lungs, stomach, and intestines highlighted. On the right, Petri dishes contain organoids—simplified, miniature versions of these organs grown in vitro. The labeled sections include:Cerebral organoids: miniature models of the brain used to study neurological development and disorders. Lung organoids: simplified lung structures used to study respiratory diseases and drug responses. Gastric organoids: models used to study gastric function and diseases such as ulcers or cancer. Intestinal organoids: models used to understand intestinal health, nutrient absorption, and intestinal diseases. This image emphasizes the use of organoids in biomedical research, offering insights into human biology and allowing for drug testing and disease modeling in a controlled laboratory environment

model accurately predicted the therapy's success 82.3% of the time. Although it was less accurate for patients who had already undergone treatment, it still performed well. Additionally, by studying these organoids, researchers have gained insights into why certain tumors resist treatment and how they can change over time, necessitating adjustments in therapy.

This research is particularly vital as pancreatic cancer is highly aggressive, and each tumor can respond differently to treatment. With this model, doctors can make more informed decisions about the best treatment options for each patient, potentially improving outcomes.

A specific prediction tool, the ESMO-MCBS (European Society for Medical Oncology-Magnitude of Clinical Benefit Scale), has been developed for hematologic malignancies. This tool quantifies the clinical benefit of new therapies in this area, providing a standardized and objective measure of their effectiveness. This further development enables a more precise assessment of the clinical benefit of cancer therapies, specifically for malignant hematopoietic system diseases.

6.10 CAR T-Cell Therapy and CRISPR CAS 9

CAR T-cell therapy is a groundbreaking treatment method for certain cancers. It was first approved in Europe in 2018 for treating patients with B-cell lymphoma and B-cell leukemia [20]. The therapy involves taking T cells from the patient's blood and genetically modifying them in a unique laboratory. These modified cells, known as CAR T cells, are then returned to the body, where they can recognize and kill specific lymphoma cells (Fig. 6.16).

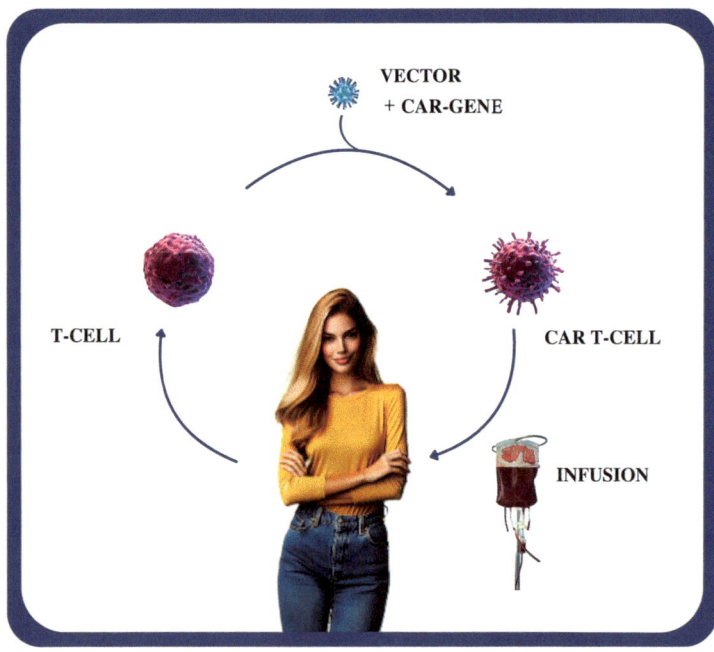

Fig. 6.16 The image illustrates the process of CAR T-cell therapy, a type of immunotherapy used to treat certain types of cancer. The breakdown of the process is as follows:Extraction of T-cells: T-cells are extracted from the patient, represented by the image of the person and the T-cell. Genetic modification: The T-cells are modified using a vector carrying the CAR gene (chimeric antigen receptor), indicated by the label "vector + CAR gene." CAR T-cell production: the modified T-cells, now called CAR T-cells, are engineered to better recognize and attack cancer cells. Infusion: The CAR T-cells are infused back into the patient, depicted by the infusion bag. This cycle represents a personalized treatment approach in which the patient's immune cells are enhanced to more effectively identify and destroy cancer cells

Innovative Concepts and Preclinical Studies

In addition to traditional applications for blood and lymphatic cancers, innovative approaches for treating solid tumors are being developed. Researchers at Columbia University have developed a concept that uses genetically modified bacteria of the probiotic strain *E. coli* Nissle 1917 [21]. These bacteria are designed to penetrate the necrotic core of tumors and attract CAR T-cells to enable a therapy similar to that used for leukemia. In preclinical studies with mice, the researchers showed that these modified bacteria colonize tumors without attacking healthy organs, making the tumor cells accessible to CAR T-cells.

Economic and Logistical Challenges

However, CAR T-cell therapy is costly. A cost analysis by the German Cancer Research Center (DKFZ) shows that producing immune cells can cost up to 305,000 Swiss francs per patient [22]. The therapy is usually only used as a last resort when other treatments have failed. To reduce costs, scientists and international partners are developing a system that will enable individualized and more cost-effective cancer therapy with CAR T-cells in the hospital.

Progress and Clinical Challenges

Despite the high costs and logistical challenges, CAR T-cell therapy is showing promising results. At a symposium of the German Society for Hematology and Medical Oncology (DGHO), new developments were presented to improve the therapy's efficiency and safety. For example, bispecific approaches such as dual CAR T-cells targeting both CD19 and CD22 and allogeneic CAR T-cells from healthy donors are being researched.

Real-World Data and Clinical Trials

Real-world data from Germany support the efficacy of CAR T-cell therapy, showing that the therapy can achieve durable remission in certain cancer patients. For example, two of the three patients who were the first to receive CAR T-cell therapy in 2010 [23] are still in remission after 10 years. The number of CAR T-cell therapies in Europe is steadily increasing, as the latest data from the European Society for Blood and Marrow Transplantation (EBMT) shows.

Application and Prospects

CAR T-cell therapy is currently mainly used for blood and lymphatic cancers [24], but clinical trials are already underway for other types of cancer, such as prostate cancer and glioblastoma. New developments, such as the MyCARe five-scoring model, should help predict the success of therapy in patients with relapsed/refractory multiple myeloma [25] and better adapt treatment to patients' individual needs.

Conclusion

CAR T-cell therapy represents a significant advance in oncology by offering new hope for patients with certain cancers. Despite the cost and logistics challenges, the results show that this therapy can be an effective and durable treatment option. With further research and technological advances, CAR T-cell therapy could become even more widely used and accessible (Fig. 6.17).

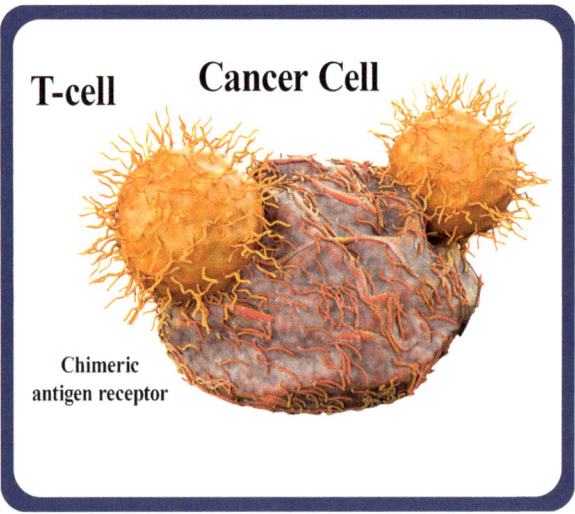

Fig. 6.17 The image illustrates the interaction between T-cells and a cancer cell in the context of CAR T-cell therapy. T-cells: the orange cells labeled as T-cells are immune cells designed to recognize and attack cancer cells. Cancer cell: the larger, irregularly shaped cell is identified as a cancer cell chimeric antigen receptor (CAR): the projections on the T-cells represent the chimeric antigen receptors. These genetically modified receptors enable T-cells to bind specifically to antigens on the surface of cancer cells, thereby enhancing the immune response. This image demonstrates how CAR T-cells recognize and destroy cancer cells, highlighting the targeted nature of this innovative cancer treatment

CRISPR CAS9 Therapy

The gene editing technology CRISPR has made significant progress in recent years and could have the potential to fundamentally change cancer treatment. CRISPR allows scientists to make precise changes to genetic material, leading to new therapeutic approaches 137,138 for various diseases, including cancer. Although CRISPR technology offers enormous potential for medical breakthroughs, such as new cancer treatments, it has also raised significant ethical and safety challenges. In 2018, Chinese scientist He Jiankui announced the birth of the first genetically modified babies: twin girls whose DNA was altered using CRISPR technology to disable the CCR5 gene linked to HIV infection. This sparked global outrage due to ethical breaches, lack of safety protocols, and inadequate oversight. He was subsequently dismissed and investigated by the Chinese authorities for possible legal violations. The incident sparked a wider debate about the ethical responsibilities of scientists, particularly those who knew of He's plans but did nothing to prevent them. The incident has led to calls for a worldwide moratorium on germline editing, as experts say the technology is not yet safe for human use. However, the controversy underlines the need for stricter global regulations and ethical guidelines to prevent premature and unsafe applications [40, 41].

CRISPR: A Game-Changer in Cancer Treatment [26, 27]

Advances in Treating Blood Disorders: CRISPR has been successfully used to correct gene mutations in mice, a development that promises to transform the landscape of human disease treatments. A significant breakthrough was the approval of CasGevi [28] in 2023, the first CRISPR-based gene therapy [29] for treating sickle cell anemia [30] and beta-thalassemia [31]. These diseases affect millions worldwide and are characterized by abnormal red blood cells. CasGevi modifies specific DNA sequences and offers a potentially curative therapy beyond traditional treatment methods. Sickle cell anemia [32], for example, causes painful crises and organ failure, while beta-thalassemia [33] requires regular blood transfusions. With CRISPR, a permanent solution may be within reach, bringing hope to millions.

Preventing T-Cell Exhaustion

Another promising approach is using CRISPR to prevent T-cell exhaustion in immunotherapies [34]. Research from Switzerland showed that inactivating a specific gene prolongs the functionality of T-cells, even in a stressful tumor environment. T-cells lacking this gene remain more active and develop better into memory T-cells, which are essential for an effective immune response. These findings could lead to more efficient cancer immunotherapies.

Personalized Therapies for Lymphomas

CRISPR also plays a crucial role in developing personalized therapies for lymphomas. Studies have shown that using CRISPR combined with Bruton tyrosine kinase inhibitors (BTK inhibitors) [35] is particularly effective for patients with specific gene alterations in the MYD88 gene. This combination activates specific autophagy processes that curb tumor growth and improve the treatment of diffuse large B-cell lymphomas (DLBCL). Such personalized approaches could significantly enhance the efficiency and precision of lymphoma treatments.

In summary, CRISPR can potentially revolutionize cancer treatment and the therapy of other severe diseases. CRISPR demonstrates how precise gene editing can be used to cure diseases, from successfully treating blood disorders like sickle cell anemia and beta-thalassemia to new approaches in immunotherapy and the treatment of rare genetic diseases. Continued research and development of this technology promise to expand medical possibilities in the coming years, providing better and more personalized treatment options for patients worldwide (Fig. 6.18).

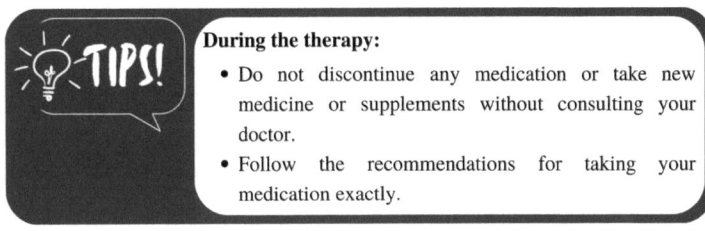

TIPS!

During the therapy:
- Do not discontinue any medication or take new medicine or supplements without consulting your doctor.
- Follow the recommendations for taking your medication exactly.

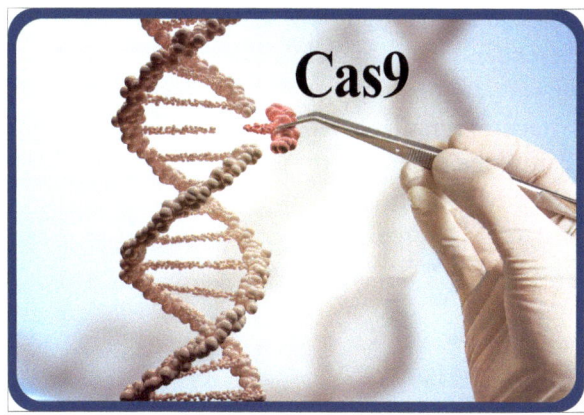

Fig. 6.18 The image depicts CRISPR-Cas9 gene editing technology. It shows a DNA double helix in which a section is precisely removed or modified, symbolized by a hand holding a pair of tweezers. Cas9: An enzyme that acts like molecular scissors and cuts the DNA at specific points. Gene editing: The representation of a precise removal or modification of DNA illustrates how CRISPR Cas9 can target and modify specific genes. This technology allows for precise genetic modifications that enable advances in genetic research, potential treatments for genetic disorders, and innovations in biotechnology

Studies

The media frequently promotes only specific findings from studies, making it difficult to assess the overall quality of the evidence. Scientific studies have varying levels of rigor, and it is critical to examine the frequency of confirmed results as well as the likelihood of replication. Unfortunately, people often accept study findings as facts even when they lack replication. In today's constant news cycle, which frequently overlooks quality, only discoveries that agree with our preexisting opinions tend to stick with us. True science does not accept conclusions at face value; instead, it asks the proper questions and uses clear techniques [36]. Scientists want outcomes that are repeatable and long-lasting. When reading about cancer research, it is critical to rely on credible sources. Original scientific papers are available in respected magazines such as PubMed. These papers go through peer review, which is an important quality assurance process in science that encourages researchers to extensively examine and assess their theories. If a theory fails to stand up, the writers must rethink their findings. Scientists must not only make statements but also answer critical questions.

The impact factor is one method for assessing the quality of a study. It represents the frequency with which other researchers mention a publication. A high impact factor indicates that the study has a significant effect. However, this statistic fluctuates according to the size of the research field. In medical research, the impact factor is especially important [36]. Scholarly journals like "Science" (impact factor 48), "Nature" (50), and the "New England Journal of Medicine" (91) are renowned for their exceptionally high impact factors.

6.11 Clinical Trials

Sometimes, all the "standard" drugs available have been given to a patient, but the cancer continues to grow despite this. It is occasionally possible to propose a new treatment, often still untested, whose side effects and mode of administration are not yet fully known, nor is there any certainty whether it will be helpful, but which we hope will work because it has already shown signs of efficacy in tests on animals or specific patients. Patients can, therefore, have access to a promising new drug, but only in a particular context: by participating in a clinical trial.

A clinical trial is part of a research study performed on human subjects. The trial aims to evaluate a new medical treatment. This is how researchers can confirm if a new drug or a new treatment is safe and effective for humans or determine whether they are more effective and/or less harmful than standard ones. However, clinical trials are not easy to run.

In a clinical study, multiple data concerning the patient, disease, and drug data are collected during treatment [37]. Analysis of this data will provide a wealth of information on the efficacy and toxicity of the new drug. This information can then be published and presented at conferences, providing the basis for tomorrow's medicine for future patients. The data collected will answer many questions, all with the aim of advancing science: which of the two treatments is better? Which is better tolerated between two drugs of identical efficacy? Is this promising drug effective in all types of cancer? Which drug dose works best? Can we avoid side effects with the same efficacy by lowering the amount? Is it worth continuing treatment over the long term, or is a short course of therapy enough? Can we predict in advance which patients will respond best to treatment or those for whom it is pointless to offer therapy because it cannot work?

Thanks to the collaboration of patients in past clinical trials, we're gradually able to answer these questions and know how to treat cancer better and better. It's thanks to them that new drugs have been discovered.

Patients make a real contribution to medical progress by participating in clinical studies. Their experience will be helpful for future generations of patients. Sometimes, they can also benefit from a brand-new drug that cannot be obtained any other way.

However, it can also happen that the new drug is less effective than expected or is responsible for unexpected side effects.

Before a trial even begins, there are steps to take to ensure safety and government authorities have to approve the trial [37]. Before a clinical trial on humans can begin, the treatment is first tried on cancer cells. If it works and protection is observed, it is tried on animals, such as mice, after ethical approval. These two tests are called preclinical tests. If these preclinical tests work, the treatment can be tried on humans after ethical approval. After cell and animal testing, only 10–20% of the products that enter the clinical trial phases will make it to market [38]. The human patient who is a subject in a clinical trial must be informed of all the necessary information about the clinical trial and give their consent. After that, the clinical trial can proceed to Phase 1.

Phase 1 Trial

The new treatment that has worked on cancer cells and animals is now given to humans. It is intended for patients who have tried all other options and hope this new intervention will help. The treatment is given at increasing doses, depending on side effects that are carefully observed and reported. The main aim of phase 1 trials is to determine the optimal dose of the treatment.

Phase 2 Trial

In Phase 2 studies, the new drug is given to a carefully selected group of patients with the same disease or cancer. This group is chosen based on their responsiveness to the drug in the Phase 1 study, ensuring that it is tested on a more homogeneous group. Moreover, Phase 2 trials allow patients to receive treatment with better-known side effects and optimal dosage. By determining the response rate, duration of response, and precise tolerability in a selected group of patients, Phase 2 studies help to provide a better understanding of the drug's efficacy. With typically several dozen patients, sometimes even more than 100 or 200, participating in a Phase 2 study, the findings can be more statistically significant, offering a greater level of confidence in the drug's effectiveness. Once the most appropriate dose has been established in the Phase 1 study, the new drug is then administered to a group of people with the same disease or cancer, for example, a group made up entirely of breast cancer patients; if it is these patients that the drug seems most effective in the Phase 1 study. This is done to obtain a more homogeneous group and to see if the drug is effective by testing it on a whole group (to be sure that its efficacy on a few individuals in a "phase 1 study" is not due to chance). Phase 2 trials enable patients to receive treatment with better-known side effects and optimal dosage. A phase 2 study aims to determine the response rate, duration of response, and precise tolerability in a selected group of patients. For a Phase 2 study, several dozen patients are generally required (sometimes more than 100 or 200).

Phase 3 Trial

If the Phase 2 trial shows a positive response, the new treatment will be compared to the best available treatment. The Phase 3 trial aims to see if the new treatment is better than the gold standard treatment or its equivalent but without worse side effects. To make an observable difference, researchers need many human subjects. Some subjects will receive the new treatment, while others will receive the existing gold standard. If there is no "gold standard," then the new treatment is compared to the outcome from placebo or best supportive care. The treatment is assigned to subjects randomly by a computer, so the subjects cannot choose which treatment they will get.

As this shows, getting a new treatment approved involves many tests and trials. Whatever your reason for participating in a clinical trial, know that you have a role to play in scientific discovery and can help more people combat deadly diseases like cancer. You will have an active role in your health care.

To ensure future success, researchers are looking for new ways to design more efficient clinical trials by using artificial intelligence (AI). With the help of AI, patients in clinical trials are chosen based on who is most likely to respond positively to the drugs being studied. This could help all cancer patients be more confident in participating in clinical trials and help advance cancer treatment [39].

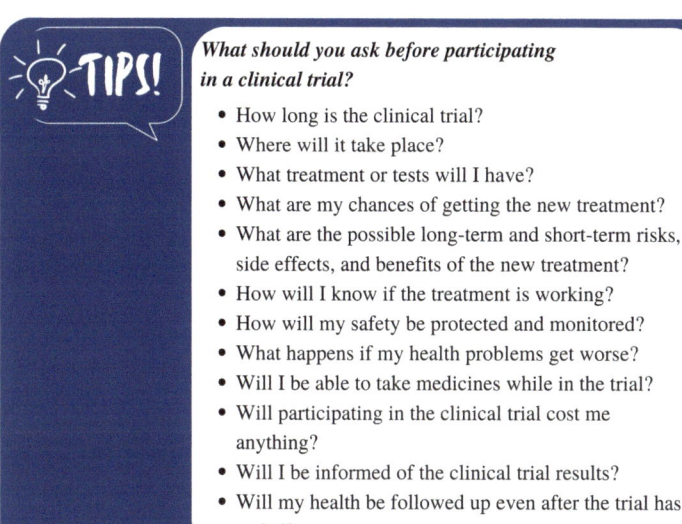

What should you ask before participating in a clinical trial?

- How long is the clinical trial?
- Where will it take place?
- What treatment or tests will I have?
- What are my chances of getting the new treatment?
- What are the possible long-term and short-term risks, side effects, and benefits of the new treatment?
- How will I know if the treatment is working?
- How will my safety be protected and monitored?
- What happens if my health problems get worse?
- Will I be able to take medicines while in the trial?
- Will participating in the clinical trial cost me anything?
- Will I be informed of the clinical trial results?
- Will my health be followed up even after the trial has ended?

6.12 What Happens When Cancer Comes Back?

When cancer comes back after being treated, it can be a scary experience. In oncology, *relapse* is when cancer returns after *remission*. A return of any sign of cancer is referred to as a relapse. Remission means all signs of cancer have disappeared after treatment. Cancer can relapse in two ways: the tumor can grow in the same place it was before *(local recurrence)*, or new tumors can form in other body parts *(distant metastases)* (Fig. 6.19). If your cancer comes back, you may need to have various tests to identify an efficient treatment. Treatment options for relapsed cancer may be like those used before. Still, sometimes options are limited due to the health condition of the patient.

As any cancer survivor knows, a cancer diagnosis is never the end of the story. Even after successful treatment, the specter of recurrence looms large. For example, in colorectal cancer, distant metastases can occur in the liver, even if there is no longer a tumor in the intestine. However, it is still colorectal cancer, not liver cancer, since those cancer cells in the liver are colorectal cancer cells, not liver cells. If the cancer returns, the diagnostic marathon begins anew. Additionally, imaging procedures such as X-rays and CT scans are performed again, and sometimes, tissue must be removed once more via biopsy to examine the new tumor.

The recurrence therapy can only be chosen after an exact diagnosis is made. The same treatment options may be available for the initial disease (such as

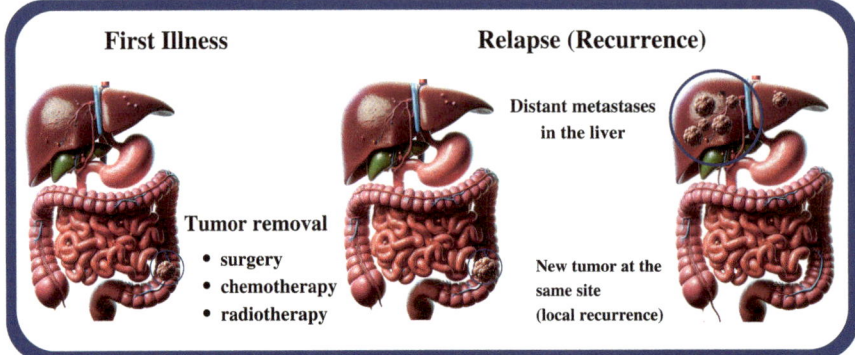

First Illness   Relapse (Recurrence)

Distant metastases
in the liver

Tumor removal
• surgery
• chemotherapy
• radiotherapy

New tumor at the
same site
(local recurrence)

Fig. 6.19 Two ways a relapse can occur. A new tumor can develop again at the site of the initial tumor, called a local recurrence. Metastases called distant metastases can also form elsewhere in the body

surgery, chemotherapy, radiation therapy, and so on). Unfortunately, using particular drugs to treat a relapse is sometimes limited due to the patient's general condition and medical history. This may be the case when an affected organ has already been operated on, making surgery impossible. Or if a patient has already received the maximum dose of radiation or chemotherapy that is safe to tolerate, using the same treatment again may not be possible. When this happens, clinical trials turn out to be the best treatment option. While this news can be daunting, it is essential to remember that clinical trials are how we progress against cancer—by testing new treatments and finding out what works best. So, if you face a recurrence, please do not despair—options are still available.

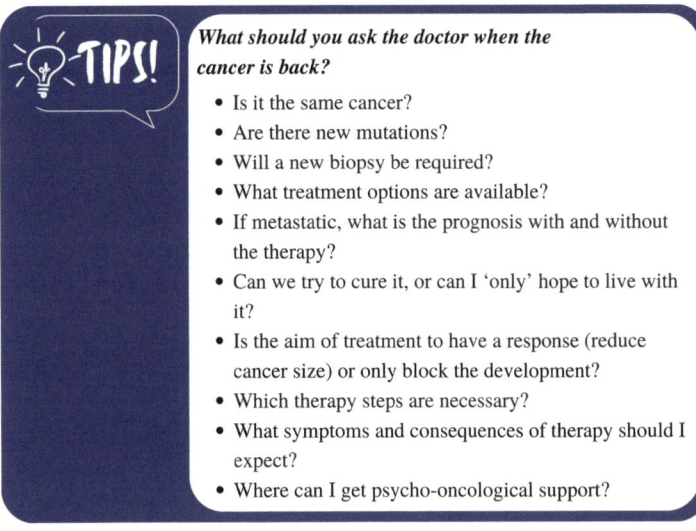

TIPS!

What should you ask the doctor when the cancer is back?

• Is it the same cancer?
• Are there new mutations?
• Will a new biopsy be required?
• What treatment options are available?
• If metastatic, what is the prognosis with and without the therapy?
• Can we try to cure it, or can I 'only' hope to live with it?
• Is the aim of treatment to have a response (reduce cancer size) or only block the development?
• Which therapy steps are necessary?
• What symptoms and consequences of therapy should I expect?
• Where can I get psycho-oncological support?

6.13 Therapy Goals in the Event of a Relapse

Therapy goals for cancer relapse include removing all detectable cancer cells, reducing the risk of cancer returning, and improving quality of life. The first step in effective therapy is to *determine the type of relapse.* This is done by looking at the original tumor site, reviewing treatment records, and performing new scans and blood tests. Treatment for cancer relapse may include surgery, radiation therapy, chemotherapy, targeted therapy, or immunotherapy. Several factors will determine the type of treatment that is chosen, including the type and stage of cancer, age, overall health, and whether the cancer has responded to treatment in the past. With advances in treatment options, more people can achieve remission from cancer than ever before. However, it is essential to remember that there is no guaranteed cure for cancer. Unfortunately, cancer can always come back. It is important to consider the previous response of cancer to treatment while choosing a new treatment. If a relapse occurs more than a year after the successful completion of a chemotherapy regimen, a similar treatment may be considered. However, in the case of a relapse just after the completion of chemotherapy treatment, resuming the same treatment is not advisable as the cancer cells may have become resistant to the treatment. It is important to choose a different treatment in such cases to prevent further progression of the disease. As such, it is crucial to *follow up* with your healthcare team and stay on top of your health so that you can catch any early signs of recurrence and start treatment right away if necessary. If a relapse occurs more than a year after a successful chemotherapy treatment, there is a chance that the same treatment could be effective again. However, if the relapse occurs immediately after the end of chemotherapy, continuing the same treatment is unlikely to be effective. This is because the cancer cells have already become resistant to the treatment. Therefore, it is important to work with your doctor to determine the best course of action based on the timing of your cancer relapse.

6.14 What to Do in Case of Hopelessness and Therapy Fatigue?

When cancer comes back after treatment, it can be a challenging and overwhelming experience for anyone. But it's crucial to understand that relapse is a possibility that may occur, and it's essential to be prepared for it. Along with disappointment and frustration, one may feel anxious and hopeless. However, various ways exist to manage these emotions and regain control. At the time of diagnosis, psychiatric or psychological support and integrative medicine can be very helpful in managing anxiety and stress, boosting self-esteem, and improving the quality of life. These therapies can help the patient take part in therapeutic decisions and develop a sense of control over the situation. Seeking help for mood problems is essential for managing these symptoms and ensuring that cancer is treated correctly. It's critical to remember that psychiatrists specialize in treating mental health issues, including depression and anxiety. These conditions can significantly impact sleep quality, appetite, and overall well-being, indirectly affecting cancer treatment. Seeking help for mood problems is essential for managing these symptoms and ensuring that cancer is treated correctly. So, it's necessary to be confident and open to seeking help to manage the emotional distress of cancer recurrence.

Summary

Following a cancer diagnosis, doctors come together on tumor boards to determine the best treatment based on tests, experience, and studies. They make decisions regarding operations, therapies, and potential side effects. Patients can also choose palliative measures. Cancer treatment needs to be personalized, as cancer and patient responses differ. Some patients benefit from local therapies such as surgery and radiotherapy, while others may require chemotherapy. It's important for patients to thoroughly discuss all options with their doctor before starting treatment, and they have the right to decline treatment if the drawbacks outweigh the benefits. For slow-growing cancers, a wait-and-see approach may be appropriate to avoid unnecessary treatment. Early treatment is often effective, but if it doesn't work, further options or trials can be considered. Surgery, with the aid of modern technologies, is particularly effective for localized cancer. These technological advancements should bring hope and confidence to patients. Radiotherapy and systemic treatments like chemotherapy target cancer cells throughout the body. Hormone therapy, immunotherapy, and targeted therapies are other specific treatments. In cases of cancer recurrence, tests are performed to determine the best treatment to eradicate cancer cells, minimize the risk of recurrence, and enhance quality of life. Support in cases of hopelessness or treatment fatigue can be provided through psychiatric care, psychological help, and integrative medicine.

References

1. Operationen: Ein Überblick für Krebspatienten. https://www.krebsinformation-sdienst.de/operation. Accessed 05 July 2024.
2. Abbasi H, Kouchak M, Mirveis Z, et al. What we need to know about liposomes as drug nanocarriers: an updated review. Adv Pharm Bull. 2022;13(1):7. https://doi.org/10.34172/apb.2023.009.
3. Elsharkasy OM, Nordin JZ, Hagey DW, et al. Extracellular vesicles as drug delivery systems: why and how? Adv Drug Deliv Rev. 2020;159:332–43. https://doi.org/10.1016/j.addr.2020.04.004.
4. Dang XTT, Kavishka JM, Zhang DX, Pirisinu M, Le MTN. Extracellular vesicles as an efficient and versatile system for drug delivery. Cells. 2020;9(10):2191. https://doi.org/10.3390/cells9102191.
5. Sabani B, Brand M, Albert I, et al. A novel surface functionalization platform to prime extracellular vesicles for targeted therapy and diagnostic imaging. Nanomed Nanotechnol Biol Med. 2023;47:102607. https://doi.org/10.1016/j.nano.2022.102607.
6. Fu Z, Li S, Han S, Shi C, Zhang Y. Antibody drug conjugate: the "biological missile" for targeted cancer therapy. Signal Transduct Target Ther. 2022;7(1):1–25. https://doi.org/10.1038/s41392-022-00947-7.
7. Gogia P, Ashraf H, Bhasin S, Xu Y. Antibody-drug conjugates: a review of approved drugs and their clinical level of evidence. Cancers. 2023;15(15):3886. https://doi.org/10.3390/cancers15153886.
8. Król M, Pawłowski KM, Majchrzak K, Szyszko K, Motyl T. Why chemotherapy can fail? Pol J Vet Sci. 2010;13(2):399–406.
9. Ramón y Cajal S, et al. Clinical implications of intratumor heterogeneity: challenges and opportunities. J Mol Med. 2020;98(2):161–77. https://doi.org/10.1007/s00109-020-01874-2.
10. Immuntherapie gegen Krebs. June 4, 2024. https://www.krebsinformationsdienst.de/immuntherapie. Accessed 05 July 2024.
11. Tan CL, et al. Prediction of tumor-reactive T cell receptors from scRNA-Seq data for personalized T cell therapy. Nat Biotechnol. 2024; https://doi.org/10.1038/s41587-024-02161-y.
12. Specialist Clinics of Australia. Why is the skin cancer rate higher in Australia?. Specialist Clinics of Australia. https://specialistaustralia.com.au/why-is-the-skin-cancer-rate-higher-in-australia/. Accessed 05 July 2024.
13. Zielgerichtete Therapie: Das Tumorwachstum punktgenau hemmen. https://www.krebsinformationsdienst.de/zielgerichtete-krebstherapie. Accessed 05 July 2024.
14. Cuzick J, et al. Effect of baseline oestradiol serum concentration on anastrozole efficacy for preventing breast cancer: a case-control study of the IBIS-II trial.

Lancet   Oncol.   2024;25(1):108–16.   https://doi.org/10.1016/
S1470-2045(23)00578-8.

15. Cuzick J, et al. Use of anastrozole for breast cancer prevention (IBIS-II): long-term results of a randomised controlled trial. Lancet. 2020;395(10218):117–22. https://doi.org/10.1016/S0140-6736(19)32955-1.

16. Wurm AA, et al. Signaling-induced systematic repression of miRNAs uncovers cancer vulnerabilities and targeted therapy sensitivity. Cell Rep Med. 2023;4(10):101200. https://doi.org/10.1016/j.xcrm.2023.101200.

17. Ärzteblatt, D. Ä. G., Redaktion Deutsches. „Die RNA-Technologie könnte das Versprechen der Gentherapie erfüllen". Deutsches Ärzteblatt. https://www.aerzteblatt.de/nachrichten/142970/Die-RNA-Technologie-koennte-das-Versprechen-der-Gentherapie-erfuellen. Accessed 04 July 2024.

18. Ärzteblatt, D. Ä. G., Redaktion Deutsches. mRNA-Impfung senkt Rezidivrisiko bei Bauchspeicheldrüsenkrebs. Deutsches Ärzteblatt. https://www.aerzteblatt.de/nachrichten/150685/mRNA-Impfung-senkt-Rezidivrisiko-bei-Bauchspeicheldruesenkrebs. Accessed 04 July 2024.

19. Weinberg RA. The biology of cancer. 2nd ed. New York: W.W. Norton & Company; 2013. https://doi.org/10.1201/9780429258794.

20. Ärzteblatt, D. Ä. G., Redaktion Deutsches. Onkologie: Kompetenznetz gibt Broschüre zur CAR-T-Zell-Therapie heraus. Deutsches Ärzteblatt. https://www.aerzteblatt.de/archiv/217571/Onkologie-Kompetenznetz-gibt-Broschuere-zur-CAR-T-Zell-Therapie-heraus. Accessed 04 July 2024.

21. Vincent RL, et al. Probiotic-guided CAR T-cells for solid tumor targeting. Science. 2023;382(6667):211–8. https://doi.org/10.1126/science.add7034.

22. Ran T, et al. Cost of decentralized CAR T-cell production in an academic non-profit setting. Int J Cancer. 2020;147(12):3438–45. https://doi.org/10.1002/ijc.33156.

23. Melenhorst JJ, et al. Decade-long leukaemia remissions with persistence of CD4+ CAR T cells. Nature. 2022;602(7897):503–9. https://doi.org/10.1038/s41586-021-04390-6.

24. Passweg JR, et al. Hematopoietic cell transplantation and cellular therapies in Europe 2022: CAR-T activity continues to grow; transplant activity has slowed. Bone Marrow Transplant. 2024;59(6):803–12. https://doi.org/10.1038/s41409-024-02248-9.

25. Gagelmann N, et al. Development and validation of a prediction model of outcome after B-cell maturation antigen-directed CAR T-cell therapy in relapsed/refractory multiple myeloma. J Clin Oncol. 2024;42(14):1665–75. https://doi.org/10.1200/JCO.23.02232.

26. Commissioner, O. of the. FDA approves first gene therapies to treat patients with sickle cell disease. FDA. https://www.fda.gov/news-events/press-announcements/fda-approves-first-gene-therapies-treat-patients-sickle-cell-disease. Accessed 04 July 2024.

27. MHRA authorises world-first gene therapy that aims to cure sickle-cell disease and transfusion-dependent β-thalassemia. GOV.UK. https://www.gov.uk/government/news/mhra-authorises-world-first-gene-therapy-that-aims-to-cure-sickle-cell-disease-and-transfusion-dependent-thalassemia. Accessed 04 July 2024.

28. Casgevy | European Medicines Agency. https://www.ema.europa.eu/en/medicines/human/EPAR/casgevy. Accessed 04 July 2024.

29. Borhade MB, Patel P, Kondamudi NP. Sickle cell crisis. StatPearls Publishing; 2024.

30. Kanter J, et al. Biologic and clinical efficacy of LentiGlobin for sickle cell disease. N Engl J Med. 2022;386(7):617–28. https://doi.org/10.1056/NEJMoa2117175.

31. Sickle Cell Disease. https://www.hopkinsmedicine.org/health/conditions-and-diseases/sickle-cell-disease. Accessed 04 July 2024.

32. Goyal S, et al. Acute myeloid leukemia case after gene therapy for sickle cell disease. N Engl J Med. 2022;386(2):138–47. https://doi.org/10.1056/NEJMoa2109167.

33. Casgevy and Lyfgenia: two gene therapies approved for sickle cell disease > News > Yale Medicine. https://www.yalemedicine.org/news/gene-therapies-sickle-cell-disease. Accessed 04 July 2024.

34. Trefny MP, et al. Deletion of SNX9 alleviates CD8 T cell exhaustion for effective cellular cancer immunotherapy. Nat Commun. 2023;14(1):86. https://doi.org/10.1038/s41467-022-35583-w.

35. Phelan JD, et al. Response to Bruton's tyrosine kinase inhibitors in aggressive lymphomas linked to chronic selective autophagy. Cancer Cell. 2024;42(2):238–252.e9. https://doi.org/10.1016/j.ccell.2023.12.019.

36. Heikenwälder H, Heikenwälder M. Der moderne Krebs—Lifestyle und Umweltfaktoren als Risiko. Berlin/Heidelberg: Springer Berlin Heidelberg; 2023. https://doi.org/10.1007/978-3-662-66576-3.

37. What are clinical trials and studies? National Institute on Aging. https://www.nia.nih.gov/health/clinical-trials-and-studies/what-are-clinical-trials-and-studies. Accessed 05 July 2024.

38. Sun D, et al. Why 90% of clinical drug development fails and how to improve it? Acta Pharm Sin B. 2022;12(7):3049–62. https://doi.org/10.1016/j.apsb.2022.02.002.

39. Artificial Intelligence gives cancer research a boost. 2019. https://www.youtube.com/watch?v=vhUu5vwYUak. Accessed 05 July 2024.

40. Jon Cohen, Did CRISPR help—or harm—the first-ever gene-edited babies?, 1 Aug. 2019, https://doi.org/10.1126/science.aay9569

41. David, C., Heidi Ledford, Genome-edited baby claim provokes international outcry, Nature. 2018;563:607–608. https://doi.org/10.1038/d41586-018-07545-0

7

I Survived; What's Next?

Contents

Abstract This chapter explores the challenges and experiences that cancer survivors face after successful treatment and during the survivorship phase, including physical side effects, emotional distress, fear of recurrence, and adapting to a new normal. We will discuss survivorship plans and stress the importance of regular follow-ups to detect cancer recurrence and other serious illnesses, providing a sense of security. Additionally, we will address the potential long-term side effects of cancer treatments, including fertility issues, and the psychological effects such as PTSD, and coping strategies for these challenges. We will also highlight the role of integrative medical therapies, spiritual support, and physical activity in recovery and restoring a healthy lifestyle. Finally, we will emphasize the importance of positive thinking, goal setting, and regaining self-confidence and control.

After cancer treatment, survivors often face unpleasant side effects. However, completing treatment does not mean that the challenge of cancer is over. There are numerous issues that cancer survivors encounter. The reassuring news is that there are many resources available. In the twenty-first century, more people have survived cancer than ever before. Therefore, we will also discuss cancer survivorship plans, the long-term effects of cancer treatments, and the psychological impact of cancer. It's important for cancer survivors to know that they are not alone and that there are numerous effective ways to deal with the challenges after treatment.

You were diagnosed with cancer, went through the therapy, and are now in remission with no signs of cancer. So, now, what is next? In this chapter, we will talk about cancer survivorship plans, the late effects of cancer treatments, and the late psychological effect of cancer. There are often some unpleasant side effects after cancer therapy. When the treatment ends, it does not automatically mean the challenge of cancer is over. There are many issues a cancer survivor faces. The good news is that resources are available.

7.1   What Does "Cancer Survivor" Mean?

A cancer survivor is someone who fulfills one of these three main criteria. First, the patient successfully passed the diagnosis period and reached the end of the initial treatment. Second, they are now transitioning from treatment to extended survival. Third, they are already long-term survivors. In the following sections, we will focus on long-term survival and cover medical, emotional, and lifestyle topics that improve our health. The frequency and type of follow-up required to detect cancer recurrence varies depending on the individual's cancer. If a cancer is aggressive and has a high risk of recurrence, follow-up appointments will be more frequent compared to a slow-progressing cancer with a low risk of recurrence. For some cancers, patients are seen every 3 months, while others require annual appointments. Generally, the risk of relapse is higher in the first few years after treatment. As time passes, the interval between two consultations may become less frequent. The purpose of posttreatment follow-up is to diagnose asymptomatic relapses early, which can be cured or better managed with early treatment to extend the patient's life. However, we only seek to detect relapses where early treatment can bring a benefit. We do not aim to detect relapses where early treatment does not change the patient's prognosis and where there is no benefit in proposing early treatment.

Are Cancer Survivorship Plans Necessary?

Nowadays, cancer specialists focus on therapy and the medical issues that cancer survivors face. With the help of ***survivorship plans,*** physicians can ***diagnose cancer recurrence*** and secondary cancers early enough to counteract them. This kind of ***regular screening*** [1] also helps medical staff find other serious diseases and helps us to maintain a healthy lifestyle. Furthermore, research strongly suggests that these plans help survivors adopt a positive mindset and feel empowered. The goal of survivorship plans is that the doctor not only monitors the cancer patients during the treatment but also continues to do so for several years after the end of the treatment. Thus, ***follow-ups*** are a must.

They enable early treatment of a local recurrence, which can often lead to a cure. Unfortunately, the same cannot be said for the early discovery of distant recurrence. In such cases, the disease can often be cured, but only in exceptional cases. For example, early detection of metastases is useful in colorectal cancer follow-up. Regular examinations such as scans and blood tests are crucial in detecting the presence of early metastases. Clinical studies have shown that early intervention at the level of metastases changes the type of treatment proposed, enabling patients to live longer and sometimes even be cured. It is, therefore, imperative to follow up regularly to detect and treat any metastases as early as possible.

7.2   Possible Late Effects of Cancer Treatment

Although there are targeted cancer therapies that save lives, there are also side effects that harm us. Some side effects may be long-term effects. And when these side effects persist even years after cancer treatment ends, they are called late effects. In this section, we will learn about the possible late effects of cancer treatment. However, it should be mentioned that most of those who are treated for cancer do not experience these late effects. Nevertheless, knowing these consequences and talking about them with your doctor is important. Certain medical treatments, including anthracycline chemotherapy, can make patients physically fragile and require special attention. Patients undergoing such treatments will be advised to avoid other risk factors for heart problems, maintain good blood pressure, quit smoking, exercise, and take good care of their feet if the nerves responsible for sensing vibrations are weakened. They will also be encouraged to avoid behaviors that could worsen their situation, such as drinking alcohol or not taking vitamin B12 supplements if they are vegetarian. For specific surgical treatments, like kidney removal for cancer, patients will need to take extra care of their remaining kidney and avoid drugs, such as anti-inflammatory drugs, that could damage it whenever possible. With intense radiotherapy, there may be a slightly higher risk of cancer in the treated area. Therefore, patients who received radiotherapy near their breasts in their youth will need close monitoring, starting well before the recommended screening age for the general population. They will receive special attention to monitor any potential late complications.

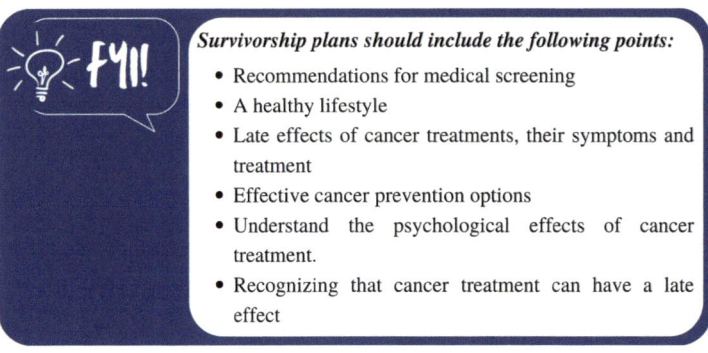

Survivorship plans should include the following points:
- Recommendations for medical screening
- A healthy lifestyle
- Late effects of cancer treatments, their symptoms and treatment
- Effective cancer prevention options
- Understand the psychological effects of cancer treatment.
- Recognizing that cancer treatment can have a late effect

Fertility Issues

Treatments like radical surgery (surgery and radiotherapy) for bladder or colon cancer or chemotherapies for other cancer types can cause *temporary or permanent infertility*. In men, the testis produces the sperm required for reproduction. Today, there are methods of preserving fertility for men, including sperm banking. Cancer treatment can also affect women's fertility. These effects include having to surgically remove organs that play a crucial role in women's reproductive cycle. Moreover, the use of chemotherapy, cancer drugs, and radiation could impact the reproductive area, the pituitary gland in the brain, or even the whole body. Cancer treatment can also cause *premature menopause* in women, which prevents pregnancy. For women, there are various fertility preservation procedures that can be discussed in "fertility boards." You can consult with a specialist, especially if you're likely to experience premature menopause due to chemotherapy or surgery. Options such as oocyte retrieval or removal of ovarian fragments are considered on a case-by-case basis. And the best part? You can consider these procedures as long as there's a reasonable delay before starting treatment and your life isn't in danger. So, don't let infertility be a side effect of treatment. Talk to your doctor today and find out how you can preserve your fertility!

Late Psychological Effects of Cancer

After a cancer diagnosis, a patient's daily life changes. Of course, the main goal of someone dealing with cancer is to eradicate that disease as fast as possible. Yet, a marathon of diagnostic tests and scans can feel emotionally and physically draining, which is natural. For patients focused only on survival, the further long-term emotional effects that arise from the disease may be regarded as secondary. It is totally understandable that survival suddenly becomes the patient's main goal in this case. Therefore, the patient may be a little concerned about the side effects of treatment. However, some patients are more afraid of those side effects than cancer itself. But that is also completely understandable, as every patient will ultimately react differently to such a life-altering disease [1].

The course of treatment can feel like a restless race where the patient keeps going, going, and going until the treatment is finally finished. Naturally, such an experience is exhausting. All the patient can think about is, "I want my normal life back." When treatment ends, there is an initial feeling of euphoria over finishing treatment, but when that feeling wears off, the patient may start to feel depressed. During the treatment, the doctors, nurses, and the other patients become like family members. However, when it is finished, the loneliness the patient feels can be challenging. At first, this may seem illogical, but nearly everyone treated for cancer goes through it. Therefore, ***do not suffer in silence***.

Sometimes, psychological help is needed. The doctors and family members who aid the patient's progression through cancer treatment form a support network that can help, even after effective cancer treatment. Another way of feeling better is to compare how one feels now with how one felt at the worst moment of treatment and not to how one was in their normal previous life. The worst day was probably day one. But then patients realize that they actually got better, and this gives them hope.

As cancer patients complete their treatment, it's normal for them to have mixed emotions. Some may feel joy at completing the treatment, while others may feel anxious about the future. However, it's important to note that these emotions are a natural, reactionary phenomenon, and there are ways to manage them constructively. After the treatment ends, the disease remains in the patient's mind. It takes time for them to come to terms with it and for it not to be a constant presence in their daily life. Each person reacts differently and gradually moves on to focus on other things, so the disease is no longer always at the forefront of their thoughts. To ensure that the disease is no longer always in the foreground, each person will react in their own way. Some may resume or start activities in a field where they can recharge their batteries and regain confidence. Others may need to surround themselves with loved ones who make them feel good. Still, others may need to isolate themselves for a while to regain their strength. It's important to remember that the time needed to catch one's breath is different for everyone, and there are no hard and fast rules to follow. While some may experience positive changes after cancer treatment, such as a newfound appreciation for life, others may feel irritable or misunderstood. They may feel disappointed that people expect them to do everything they did before the illness, even though they still feel fragile and vulnerable. Patients may also feel abandoned by the medical team, with whom they have fewer and fewer appointments. Despite completing their treatment, they may feel anxious about the risk of relapse or the difficulties of coping with the changes brought about by cancer treatment. It's important to recognize when these emotions start to disrupt daily life, interfere with sleep or appetite, prevent a positive outcome from being seen, prevent people from making plans, or are responsible for anxiety attacks. In such cases, psychiatric advice and sometimes treatment are required. There are many constructive ways to cope with the challenges that come with cancer treatment. Coping mechanisms such as integrative medicine therapies, spiritual support, the gradual resumption of regular sporting activity, and contact with nature can all be helpful during this difficult phase. By seeking psychological or psychiatric support and utilizing these coping mechanisms, patients can restore harmony to their daily lives and continue to move forward with hope and positivity.

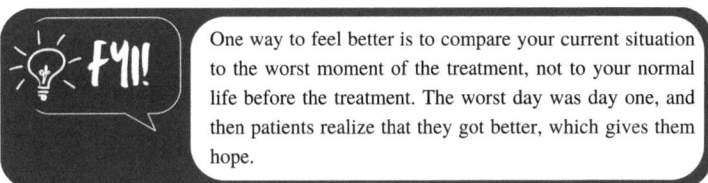

One way to feel better is to compare your current situation to the worst moment of the treatment, not to your normal life before the treatment. The worst day was day one, and then patients realize that they got better, which gives them hope.

We completely understand. It's common for patients to feel anxious to resume their normal life after an illness. It's frustrating when progress is slow, and patients tend to compare their current health status to how they were before the illness, which can make them feel worse. Instead, focusing on the worst time of the illness can help patients see the progress they've made and give them hope and motivation to continue. It's important to remember that you're doing better now than you were at the worst time, not worse than before the illness. One way to resume a normal life is to set small goals unrelated to the illness. These goals can be simple, such as taking a walk every day, reading a book, or starting a new hobby. Achieving these goals can help patients regain confidence in themselves and feel a sense of control over their lives. It's important to remember that the illness shouldn't define them, and they can still enjoy life by setting these personal goals.

Post-traumatic Stress Disorder

Post-traumatic stress disorder (PTSD) [1] is an anxiety disorder triggered by an extremely frightening or life-threatening situation in which one survives. PTSD is a recognized condition among cancer survivors, triggered by the trauma of diagnosis and treatment. Unlike PTSD resulting from singular traumatic events such as an accident, war, sexual and physical abuse, cancer-related PTSD develops over an extended period, often linked to ongoing medical interventions, uncertainty about the future, and the persistent fear of recurrence. The typical symptoms of PTSD are nightmares, flashbacks, or avoiding people, places, and things linked with the event. In addition, with PTSD, there is a strong feeling of guilt, hopelessness, and shame. Self-destructive behavior may also occur. These symptoms can develop 3 months after the traumatic event or even years later. Therefore, cancer survivors and those still struggling with the disease may suffer from PTSD. In this case, they must *get the proper treatment.* If not, PTSD can interfere with their follow-up care or personal lives. They can be at increased risk of developing other physical, mental, or social issues that lead to lost jobs and broken relationships. Finally, as with anxiety, caregivers can develop PTSD as well.

Fear of Recurrence

Recurrence means that cancer has returned after a successful treatment where no cancer cells were detectable. However, cancer is an unpredictable disease; it is sneaky and can come back. Therefore, it is more than natural for a cancer patient to be worried about a potential recurrence. But feeling worried is not a negative thing. In fact, it can motivate you to *start a healthier lifestyle,* like quitting smoking or having regular screenings.

On the other hand, worrying too much can be emotionally paralyzing. That is why it is important to *think positively*. It is crucial to understand that the psychological effects of cancer may continue even after therapy ends. Taking proactive steps, such as seeing a counselor or joining a support group, is beneficial. Other successful posttreatment strategies include talking to your doctor, family and friends, as they may have a wealth of incredible tips and knowledge to share with you.

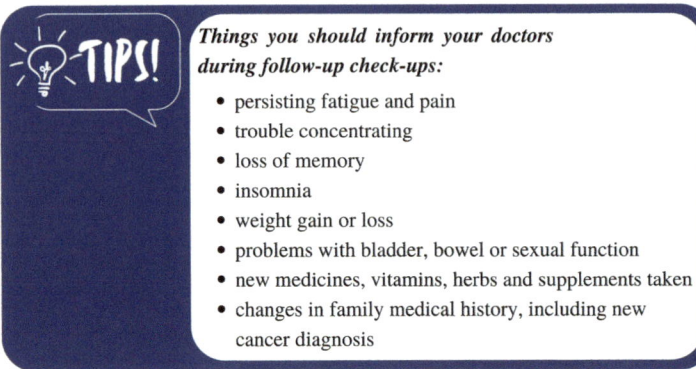

Things you should inform your doctors during follow-up check-ups:

- persisting fatigue and pain
- trouble concentrating
- loss of memory
- insomnia
- weight gain or loss
- problems with bladder, bowel or sexual function
- new medicines, vitamins, herbs and supplements taken
- changes in family medical history, including new cancer diagnosis

7.3   Follow-Up Care

Regular follow-up appointments are crucial in detecting any possible cancer recurrence and providing prompt treatment to cancer patients. The frequency and types of follow-up appointments may vary depending on the type of cancer, with more frequent check-ups required for aggressive cancers that carry a high risk of recurrence. However, follow-up appointments may be less regular for cancers with a lower risk of recurrence and slow progression [1]. The main objective of posttreatment follow-up is to identify asymptomatic relapses early, which can be cured, or where early treatment can extend the patient's life. To achieve this, doctors aim to detect only those relapses for which early treatment is beneficial and avoid screening for regressions that do not change the patient's prognosis or benefit from early treatment. For example, in the case of breast cancer, regular radiological checks of the breast, such as mammography/ultrasound or MRI, are crucial in detecting any local duplications early, which can often lead to a cure. However, screening for metastatic breast cancer relapse may not be necessary since patients cannot feel it, and it can only be treated in exceptional cases. Early treatment of asymptomatic degeneration does not increase the patient's life expectancy. However, early detection of metastases in colorectal cancer can change the type of treatment proposed and improve the patient's lifespan, sometimes even leading to a cure. Therefore, follow-up appointments may involve regular examinations, such as scans and blood tests, to detect the presence of early metastases. As we can see, follow-up appointments vary based on the type of cancer and the patient's situation. Adhering to regular follow-up appointments can help detect cancer recurrence early and provide timely treatment, leading to better patient outcomes.

7.4   Prevention

Preventing cancer is a simple and cost-effective measure that brings numerous additional benefits. Scientists find it surprising that school curriculums don't more widely integrate sports, despite the fact that daily physical activity should be an integral part of the school day, ideally starting in the morning. The World Health Organization (WHO) recommends that children aged 5–17 engage in 1 h of physical activity each day [2]. Starting the day with exercise not only promotes social equity by giving all children access to physical activity regardless of their background, but it also has a host of positive effects on academic performance. Research shows that an hour of morning exercise boosts concentration, improves grades, helps regulate body weight, lowers the risk of diabetes and back pain, reduces stress and performance pressure, and even enhances intelligence test scores [3–7]. Accordingly, 86% of the US population doesn't get enough exercise, which has serious long-term health implications [8]. Regular exercise conditions the body to perform at a higher level, making physical activity a natural part of daily life. People who exercise regularly often miss the "runner's high" when they're unable to work out due to health issues. Endorphins, our body's natural happiness hormones, cause this feeling and serve as a powerful incentive to stay active. We can minimize the impact of environmental mutations by eliminating cancer-promoting factors. Although it's impossible to prevent all mutations over a lifetime, healthy lifestyle habits can slow their progression. We can't directly control metabolism, which drives most mutations, making this particularly important.

A renowned team of cancer researchers, including Bert Vogelstein, has outlined the long-term progression of colorectal cancer from benign polyps to prevention through early detection, such as endoscopy, and estimates that we could prevent 70% of cancer cases worldwide each year [9, 10]. Prevention occurs at a variety of stages: primary prevention focuses on identifying and avoiding cancer's causes. While most studies have focused on middle-aged adults, it's equally important to investigate risk factors during childhood and adolescence in order to address them early. We could tailor future personalized prevention strategies to individual risk profiles, considering factors like preexisting conditions, hormonal contraceptive use, blood glucose and lipid levels, hormone levels, exercise habits, diet, and alcohol consumption to design optimal prevention plans.

Screening programs often exclude young people despite the potential benefits of early diagnosis. Early colonoscopy is important for high-risk individuals, even before age 45, and regular cancer screenings are critical for early detection. Early detection often leads to successful treatment of cancer. Once we detect and treat cancer, we must implement specific measures to detect recurrences early and enhance the quality of life for individuals with chronic conditions. The discovery of new biomarkers could greatly increase the number of cancers diagnosed at an early stage. Secondary prevention focuses on timely cancer detection and treatment. Tertiary prevention aims to enhance follow-up care and quality of life post-treatment.

Despite our incomplete understanding of how food, heavy metals, and other environmental chemicals contribute to cancer, it's wise to avoid unnecessary exposure to carcinogens.

Vaccinations

Vaccinations are a crucial yet often underappreciated method of cancer prevention. While their benefits may not be immediately obvious, vaccines play a vital role in protecting against specific cancers, and widespread participation in vaccination programs could even lead to the eradication of cancer-causing viruses.

For instance, many people opt for hepatitis A and B vaccinations before traveling to regions like Africa and Southeast Asia, where these viruses are common and significantly increase the risk of liver cancer. However, it's important to note that even a trip to nearby Mediterranean countries could expose travelers to these viruses through contaminated water or seafood.

The human papillomavirus (HPV) vaccine, which has been available for over 16 years, is another key preventive measure. This vaccine can protect against cervical cancer, yet the vaccination rate in Germany is alarmingly low, with less than half of girls and only 5% of boys vaccinated. Many people are either unaware of the vaccine or choose not to get it, leading to preventable cases of cervical cancer. The HPV vaccine is a critical tool in cancer prevention due to its well-documented effectiveness in preventing this cancer.

In less developed regions, such as parts of Africa and South America, cervical cancer remains one of the most common and deadly cancers, often surpassing deaths from liver and colorectal cancers combined. While wealthy countries struggle with awareness and motivation, poorer countries lack the medical infrastructure and financial resources to provide widespread vaccinations.

Liver cancer is also prevalent in developing countries due to limited access to hepatitis vaccinations. Ignoring these preventive measures might lead to severe health problems, infertility, or cancer treatments, making vaccination a cost-effective and less painful option despite the inconvenience.

Scientists are currently working on developing new vaccines, including mRNA-based vaccines designed to specifically activate the immune system against mutated proteins in tumor cells. These innovative vaccines offer hope for patients who have not responded to conventional treatments due to the complexity and variety of mutations. For certain types of cancer, these vaccines are already being tested or in use. The goal is for cured cancer patients to live free from fear of relapse and without the severe side effects of traditional therapies.

Notably, Prof. Dr. Uğur Şahin was awarded the German Cancer Prize in 2019 for his groundbreaking work on mRNA-based vaccines, which are tailored to the mutation profiles of individual cancer patients. The development of the COVID-19 mRNA vaccine was a by-product of this technology, with its primary focus being personalized cancer therapy [10].

Summary

After a cancer diagnosis and successful treatment, cancer survivors have to deal with the challenges of survival and the possible late effects of treatment. Survivorship plans help to recognize and treat relapses and secondary diseases at an early stage. Side effects and late effects, such as heart problems or infertility, require close collaboration with the medical team. Psychological sequelae, including PTSD, can affect life after treatment and often require psychological support and coping strategies. Focusing on progress and setting small, personal goals to regain self-confidence is essential. However, it's important to note that the active participation of survivors in regular follow-up appointments is crucial for the early detection of cancer recurrence and contributes significantly to better treatment outcomes.

References

1. Libov C. Cancer survival guide: how to conquer this disease and live a good life. 1st ed. Humanix Books; 2016.
2. WHO guidelines on physical activity and sedentary behaviour: at a glance; 2020. ISBN 978-92-4-001488-6. https://iris.who.int/bitstream/handle/10665/337001/9789240014886-eng.pdf

3. Ratey J, Spark J. The revolutionary new science of exercise and the brain In: Hagerman E, Series editor. Spark: the revolutionary new science of exercise and the brain. New York: Little, Brown and Co; 2008. pp ix, 294

4. Saxena M, van der Burg SH, Melief CJM, Bhardwaj N. Therapeutic cancer vaccines. Nat Rev Cancer. 2021;21(6):360–78. https://doi.org/10.1038/s41568-021-00346-0.

5. Harveson AT, Hannon JC, Brusseau TA, Podlog L, Papadopoulos C, Hall MS, Celeste E. Acute exercise and academic achievement in middle school students. Int J Environ Res Public Health. 2019;16(19):3527. https://doi.org/10.3390/ijerph16193527.

6. Barbosa A, Whiting S, Simmonds P, Scotini Moreno R, Mendes R, Breda J. Physical activity and academic achievement: an umbrella review. Int J Environ Res Public Health. 2020;17(16):5972. https://doi.org/10.3390/ijerph17165972.

7. Merriman W, González-Toro CM, Cherubini J. Physical activity in the classroom. Kappa Delta Pi Rec. 2020;56(4):164–9. https://doi.org/10.1080/00228958.2020.1813518.

8. Booth FW, Roberts CK, Thyfault JP, Ruegsegger GN, Toedebusch RG. Role of inactivity in chronic diseases: evolutionary insight and pathophysiological mechanisms. Physiol Rev. 2017;97(4):1351–402. https://doi.org/10.1152/physrev.00019.2016.

9. Song M, Vogelstein B, Giovannucci EL, Willett WC, Tomasetti C. Cancer prevention: molecular and epidemiologic consensus. Science. 2018;361(6409):1317–8. https://doi.org/10.1126/science.aau3830.

10. Ärzteblatt, D. Ä. G., Redaktion Deutsches. Ugur Sahin: Mit individualisierten Therapien gegen den Krebs. Deutsches Ärzteblatt. https://www.aerzteblatt.de/archiv/207247/Ugur-Sahin-Mit-individualisierten-Therapien-gegen-den-Krebs. Accessed 29 Aug 2024.

8

Nutrition for Cancer Prevention and Therapy

Contents

Abstract This chapter emphasizes the vital role of nutrition in both cancer prevention and treatment, highlighting how dietary choices can significantly impact health. A balanced, plant-based diet rich in fruits, vegetables, fiber, and polyphenols is linked to reduced cancer risk, while harmful foods like processed meats, alcohol, and ultra-processed products increase the likelihood of cancer. Maintaining a healthy gut microbiome and following concepts like Hara Hachi Bu—eating until 80% full—can positively influence overall well-being. The chapter also cautions against over-reliance on dietary supplements, recommending their use only under medical guidance. For cancer patients, an adapted diet can alleviate symptoms and enhance quality of life, showcasing the importance of food synergy in fostering health and recovery.

8.1   Nutrition

Eating is an integral part of our life. We eat to survive, to satisfy our needs, and for pleasure, too. Food can be found in abundance in nature and from plants and animals. Additionally, we have learned how to process and manufacture food to improve its taste and make eating convenient. We are also aware that a healthy diet helps us live a good life and, to a certain extent, helps prevent certain diseases. Understanding how food affects our bodies is what the field of nutrition is about. Nutrition is a complex topic, as the human body is incredibly nuanced and sensitive to changes in diet. While we understand how some nutrients work, many questions remain. And yet, nutritionists and other health experts have made significant progress recently.

Of course, while diet is not the only factor related to cancer, it is a vital piece of the puzzle that we continue to learn more about daily. Other important factors often discussed include water, air, and food quality. While water and air quality are easier to control in Western Europe, food quality can be harder to control. This is because many components make up food. Therefore, it can be challenging to determine which ones are good or bad for us.

In general, we should pay attention to the nutrients in our food and to the contaminants that may be present. Nutrients are essential for our bodies, and we must ensure that we get enough of them. However, too many nutrients can also be harmful, so it is crucial to find a balance. Contaminants can also cause serious health problems, so it is vital to avoid them whenever possible. By keeping these things in mind, we can ensure that we eat healthy, nutritious food to help us stay healthy and happy.

Asking if there is a cancer-fighting food is like asking if there is a superfood; the answer is clearly no. Nutrition and cancer are two broad and complicated topics. We cannot expect one single food to impact a system as vast and complex as the human body. However, that does not mean that diet plays no role in cancer. What we eat can influence our risk of developing cancer and also our chances of surviving cancer. For example, a diet high in processed meats has been linked with an increased risk of colorectal cancer.

On the other hand, a diet rich in fruits and vegetables may help to reduce the risk of developing certain types of cancer. Additionally, there is evidence that certain dietary interventions may help improve cancer treatment outcomes. There is indeed a well-known association between being overweight and the likelihood of experiencing a recurrence of breast cancer [1]. Additionally, many studies are presently being conducted to examine the influence of certain diets on the body's response to immunotherapy. While more research is needed, the potential for diet to impact cancer is clear. As we continue to learn more about the role diet plays in cancer, we may be able to develop targeted interventions that could help us prevent or treat the disease.

8.2 Cancer-Risk Food

In our fast-paced world, food manufacturing has to adapt rapidly to keep up with consumer demand. As more and more people lead busy lives, they have less time available to cook healthy meals from scratch. As a result, they eat more processed food or fast food loaded with sugar and fat. This is why those foods have become increasingly accessible. Fast food is often cheaper than healthier alternatives, making it easy to overeat without being aware of the consequences that overeating has on our health. For instance, blood sugar levels will rise rapidly when we eat too quickly and without taking proper breaks between bites. This can lead to insulin production difficulties, increasing the risk of developing type 2 diabetes [2]. While the link between diabetes and cancer is a confusing one, studies have proven that diabetic patients are more likely to experience cancer. However, the reason for this is still being researched [3].

Nevertheless, there is some good news. For example, the Japanese concept of *Hara Hachi Bu* offers a way to live healthier and potentially longer. *Hara Hachi Bu* is the belief that one should only eat until they are 80% full. This practice results in eating less overall, which can lead to health benefits like weight loss and lower blood sugar levels. So, while the link between diabetes and cancer is still being explored, there are things we can do in the meantime, like eating less to improve our health. If we follow the Japanese tradition of *Hara Hachi Bu,* it can improve our health and extend our life expectancy [4].

As for the Western diet, it is rich in processed meats, sugary drinks, and unhealthy fats. Therefore, the Western diet has been linked to increased obesity, diabetes, and cancer. Unfortunately, the Western diet has infiltrated every

Fig. 8.1 Western dietary habits are often linked to an increase in cancer risk. Japan, in particular, experienced a higher incidence of colorectal cancer as they switched from a traditional plant-based diet to a high-carbohydrate and high-fat Western diet

corner of the globe, often with disastrous consequences. Take, for instance, the case of cancer incidence in Japan. In the 1980s, the Japanese began adopting Western dietary habits, swapping healthy carbohydrates and vegetables for processed meats and sugary drinks. By 2005, the rate of colorectal cancer in Japan had reached epidemic proportions, making it the highest in the world. Fortunately, cancer rates have declined lately as more Japanese people have returned to healthier eating habits. Still, this example illustrates the potentially devastating impact that the Western diet has on global health (Fig. 8.1).

While there are many things we can do to lower our risk of developing cancer, such as maintaining a healthy weight and exercising regularly, what we eat and drink also plays a role. Studies indicate that certain foods and beverages can increase cancer risk, so we have to be aware of which ones to limit or avoid altogether. Let us discuss some of these in detail.

8.3   Alcohol

Alcohol is an everyday staple in many social gatherings and events. Many patients are often taken aback when they learn that having an apéritif at the end of the day, on top of wine or beer consumed during meals, can pose a health risk. While it may seem harmless, alcohol contains ethanol, produced when sugar and starch are fermented by yeast. When we drink alcohol, it is broken down into acetaldehyde, a chemical that can damage DNA and proteins [5]. Not only that, but ethanol also generates reactive oxygen species that oxidize and damage our DNA. At the same time, alcohol also acts as a solvent that helps environmental carcinogens penetrate the cells [6] (Fig. 8.2). In other words, alcohol consumption can impair the body's ability to break down and absorb specific vitamins it needs to stay healthy and fight cancer [7, 8].

Did you know that even a small amount of alcohol can increase your risk of cancer? Just one glass of beer, wine, or hard liquor per day, which contains 10 g of alcohol, could be enough to elevate your risk. Heavy smokers who routinely consume high-proof alcohol have a 100-fold increased chance of developing cancer of the mouth and throat compared to smokers who do not consume alcohol. Although alcohol is not commonly considered a carcinogen, its high presence can significantly damage the outermost layer of cells in the mouth and throat. This process of destruction compels the underlying cells to undergo more frequent division, hence substantially elevating the likelihood of developing cancer [9]. Alcohol functions as a carcinogen by impairing the body's defense mechanisms. It is crucial to note that alcohol is

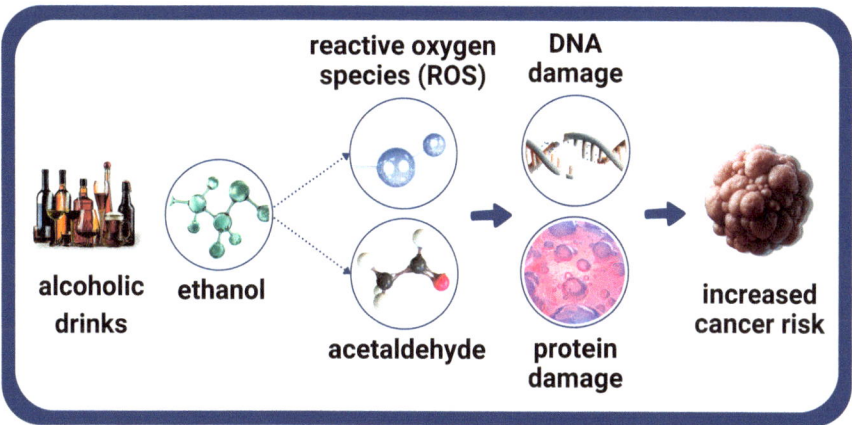

Fig. 8.2 Alcoholic drinks increase cancer risk. Ethanol in alcohol produces reactive oxygen species and acetaldehyde that damage the cells and increase cancer risk

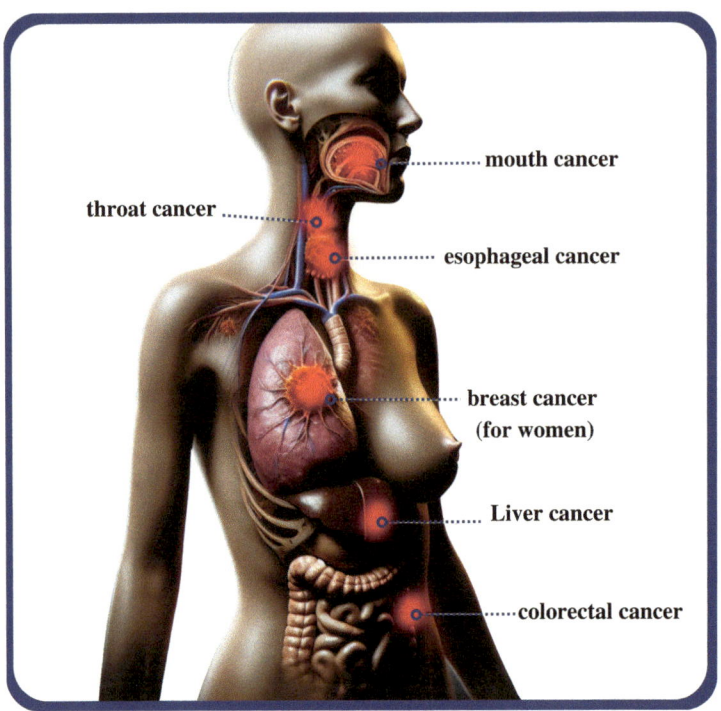

throat cancer

mouth cancer

esophageal cancer

breast cancer
(for women)

Liver cancer

colorectal cancer

Fig. 8.3 Cancers are associated with alcohol consumption. High alcohol consumption causes cell damage and increases cancer risk in certain organs

metabolized not only in the liver but also in breast tissue. Acetaldehyde, the resultant chemical, is carcinogenic and augments the susceptibility to liver and breast cancer [10].

Take care of your health and consider reducing your alcohol intake today. Figure 8.3 conveys the different cancer types associated with alcohol consumption. As we learned, heavy alcohol consumption can be harmful to our health. It can cause damage to the cells in our body and increase the risk of cancer in specific organs, such as the liver and breast. In combination with smoking, it can increase the risk of cancer of the throat, mouth, and esophagus. It's important to know how much alcohol you consume and the associated risks. Like driving a car, the risk of accidents or cancer increases with consumption. The more you drink, the greater your risk of developing alcohol-related cancer. Even a small amount of alcohol can be harmful.

According to research, the risk of cancer increases with the consumption of 10 g of alcohol a day—which is equivalent to a glass of beer, wine, or hard liquor. It's important to keep in mind that there is no "safe" amount of alcohol to drink. It's the quantity that counts, and the risks increase with consumption. Avoiding alcohol altogether may not be feasible for everyone, but it's what we should be striving to minimize our risks. Some countries, like France, have set guidelines for alcohol consumption. They recommend avoiding exceeding 10 "standard" glasses per week, no more than two per day, and planning several days each week without consuming any alcohol [11].

> ### *THREE QUESTIONS ASKED AND ANSWERED*
>
> **Can genetics affect the risk of getting cancer from alcohol?**
>
> Yes. Our genes produce enzymes that help break down alcohol. One enzyme is ADH, which converts ethanol into carcinogenic acetaldehyde. People with a superactive ADH enzyme convert ethanol into acetaldehyde more quickly and are more likely to get cancer than people with a common ADH enzyme.
>
> The second enzyme, ALDH2, transforms acetaldehyde into harmless compounds. People with a faulty ALDH2 enzyme experience unpleasant side effects when drinking alcohol due to acetaldehyde buildup. As a result, they consume less alcohol.
>
> **Can drinking red wine help to prevent cancer?**
>
> Red wine contains resveratrol, a compound found in grapes. So far, researchers have not found a link between resveratrol and cancer prevention.
>
> **Will cancer risk decrease after a person stops drinking alcohol?**
>
> Studies show that stopping alcohol consumption does not immediately reduce cancer risk. It may take years to lower the risk of those who don't drink alcohol, but it is possible.

8.4   Red Meat and Processed Meat

A large percentage of the world's population consumes meat. This, however, comes at a cost. According to a study done in 2016, 57% of the world's population consumes red meat, while pork is eaten by 36%. Though denying how good a barbecue smells is complex, we must be careful. The International Agency for Research on Cancer (IARC) classified red meat and processed meat as carcinogenic [12] (Fig. 8.4). The main difference between the two is that processed meat has undergone salting, curing, fermentation, smoking, or other processes to enhance flavor or preserve it. This does not mean that you will get cancer every time you have a burger. The IARC classifies red and processed meats as Group 2A, meaning there is a probable link between the two. In 2015, the IARC reported that eating 50 g of processed meat (e.g., a few slices of ham) daily increases the risk of colorectal cancer by 18%. For red

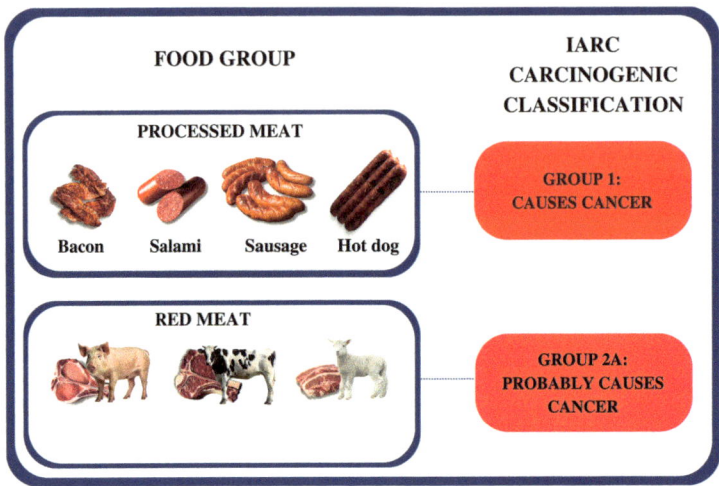

Fig. 8.4 IARC Carcinogenic Classification Groups. These categories represent the confidence of IARC that red meat and processed meat can cause cancer in humans [13]

meat, the number is slightly lower at 17%. While the chances are small, they are worth considering, especially if you already have a family history of cancer. Eating a ham sandwich every day is undoubtedly a habit worth changing.

Red meat is the unprocessed muscle meat of mammals such as cows, sheep, pigs, and goats. Beef, veal, pork, lamb, and goat meat are thus considered red meat. According to research, cooking red meat at high temperatures [14] produces carcinogenic compounds known as heterocyclic amines (HCAs) and polycyclic aromatic hydrocarbons (PAHs). These harmful chemicals can increase our risk for cancer by causing DNA damage, which accumulates over time [15] (Fig. 8.5). Therefore, it is essential to know how HCAs and PAHs are formed and how to minimize our exposure to them.

When cooked at high temperatures, the amino acids in red meat react and create HCAs. Fats and juices from the heat can also drip onto a hot surface or open fire and form PAHs. The amount of HCAs and PAHs produced depends upon the cooking method we use and the "doneness" of the meat (i.e., rare, medium, or well done). Well-done and grilled meat produces high concentrations of HCAs, while smoked meat has high concentrations of PAH.

So, what can you do to protect yourself? The best way to reduce your exposure to HCAs and PAHs is to choose lean cuts of meat and cook them by using methods that do not involve direct contact with flames or smoke. Steaming, baking, boiling, or braising are all practical options. When grilling, try to avoid charring or burning the meat. These simple precautions allow you to enjoy your barbecue without risking your health.

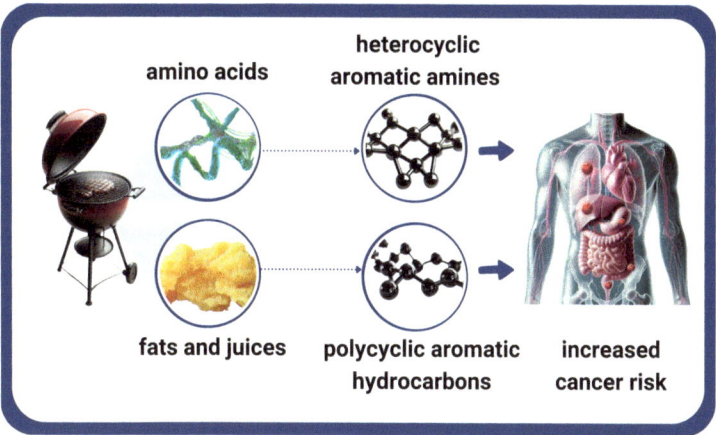

Fig. 8.5 Grilling causes the formation of carcinogenic substances. These carcinogens, HAA and PAH, cause an increase in the risk of developing colorectal, pancreatic, prostate, and other cancers

It is no secret that ***processed meats*** are packed with unhealthy ingredients. From nitrites to sodium, these foods are loaded with preservatives that can harm our bodies. However, what exactly do these preservatives do to us? When we eat processed meat, we ingest nitrites, which are preservatives that help keep the meat from spoiling. However, these nitrites can also react with the bacteria in our intestines to form nitrosamines [16]. These nitrosamines can then cause DNA mutations, which can lead to cancer. According to the World Health Organization, eating just 50 g of processed meat each day can increase your risk of developing colorectal cancer [17]. Furthermore, there is also a correlation between processed meat consumption and stomach cancer [12]. So next time you reach for that bacon sandwich, remember that it might be best to choose something else.

- Cook the meat at a lower temperature and with less heat.
- Avoid allowing flames to come into direct contact with the meat.
- Keep cooking surfaces clean by removing any charred residues.
- Marinate meat in an acidic solution to prevent cancer-causing substances from forming.
- Limit the amount of red meat and processed meat you eat and replace it with chicken, fatty fish, eggs, vegetables, and legumes.
- Eat garlic with processed meat to prevent the formation of nitrosamines.

8.5   Refined Sugar

Sugar is a complex topic. On the one hand, sugar is an essential part of our diet. It helps provide energy for our cells and brain. It can also be found naturally in fruits and vegetables. However, a significant amount of sugar we consume today comes from adding refined sugars to highly processed foods and soft drinks [18]. This type of sugar can adversely affect our health. Negative effects include tooth decay and weight gain. Therefore, it is crucial to be aware of the different types of sugar and how they impact our health. Sugar is often demonized as carcinogenic (cancer-causing), but that is not the case. Sugar is a chemically reactive compound, but it is not carcinogenic.

However, consuming too much refined sugar does contribute to obesity and diabetes, both of which are risk factors for cancer [19, 20]. In addition, the cancer cells in nearly a third of common cancers, such as breast and colon cancer, have insulin receptors on their surface (Fig. 8.6). Therefore, consuming too much sugar causes the body to produce more insulin, which binds to these receptors and becomes a catalyst to fuel cancer development [22]. So, while sugar may not cause cancer directly, it can indirectly increase your risk. Therefore, it is best to enjoy sugar in moderation.

Sugar consumption is a major contributor to various health issues, including obesity, diabetes, and cancer. When we consume sugar, our muscle cells absorb most of it and store it as glycogen, which provides energy for physical activity. The liver also stores excess sugar as glycogen, while adipose tissue

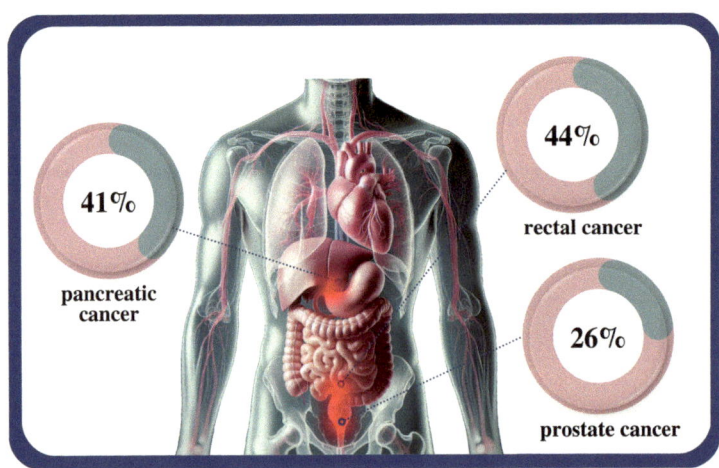

Fig. 8.6 Refined sugar: Cancer-risk. Consuming a high amount of sugar increases the risk of prostate cancer by 26%, rectal cancer by 44%, and pancreatic cancer by 41% [21]

converts it into fat, which can delay fat burning. Continuous sugar intake can lead to fat accumulation in liver cells, resulting in a potentially dangerous fatty liver, often accompanied by inflammation that promotes cancer. Over time, high sugar consumption can cause insulin resistance, where cells no longer respond adequately to the metabolic hormone insulin. This leads to hyperglycemia (elevated blood sugar) and raises the risk of liver cancer. Obesity can also contribute to insulin resistance, which is caused by free fatty acids released from fat cells, with visceral fat being the most problematic.

High sugar consumption affects all body cells, including damaged and potentially cancerous ones. Many cancer cells prefer sugar as their primary energy source, benefiting greatly from an excess of it. The Warburg effect explains that cancer cells have a much higher sugar demand than healthy cells. Healthy cells use glycolysis to convert sugar into pyruvate, which the mitochondria then further metabolize into carbon dioxide through the citric acid cycle. However, cancer cells continue to rely on glycolysis, even when oxygen levels are sufficient. This process is energetically inefficient, making cancer cells heavily dependent on a constant sugar supply. However, it's a misconception to believe that cutting out sugar and carbohydrates entirely can "starve" cancer. Since unintentional weight loss is a common and life-threatening side effect of many cancers, a strict diet could actually shorten a patient's life.

It's essential to rethink our eating habits to avoid obesity and related health risks. Since our bodies don't require industrial sugar, the easiest way to reduce sugar intake is to avoid refined sugar, processed sugary foods, and sweetened beverages. Additionally, paying attention to the quantity and quality of fats in our diet is crucial for maintaining a healthy, balanced diet. We must obtain high-quality fats, particularly essential fatty acids like omega-3, through our diet. Occasionally consuming small amounts of sugar won't dramatically impact our health, especially if we get enough exercise and eat a generally healthy diet. Nevertheless, it's wise to consume sugar in moderation to minimize long-term health risks.

8.6   Salt

Salt is not just a delicious seasoning that makes food more palatable. It is also essential for normal bodily functions. Our bodies need salt to maintain fluid balance, absorb nutrients, and transmit nerve impulses. This is why salt is indispensable. However, we unknowingly consume much more salt than health experts recommend. They recommend a daily salt intake as low as one teaspoon containing 2.4 g and up to 7 g of sodium [23]. However, since salt is often used in food processing as a preservative, over 70% of the salt we take in is hidden in packaged and prepared food [24], often in very high amounts. Eating too much salt damages the stomach lining and attacks the protective mucus layer in the intestine. As a result, the protective intestinal bacteria are destroyed. This may increase the risk of inflammation and *Helicobacter pylori* infection, two risk factors for cancer [25]. If you buy processed foods, first try to find out how much salt they contain. And the next time you reach for the salt shaker, think twice about how much you use.

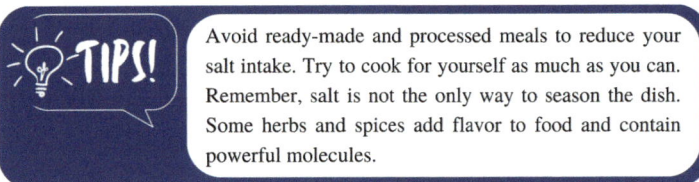

TIPS! Avoid ready-made and processed meals to reduce your salt intake. Try to cook for yourself as much as you can. Remember, salt is not the only way to season the dish. Some herbs and spices add flavor to food and contain powerful molecules.

8.7   Fats and Obesity

Over the past few decades, research has revealed a troubling reality: Obesity is not just a cosmetic concern but a major risk factor for various cancers. The global prevalence of obesity has surged, bringing severe health implications [26].

Obesity is widespread, and the statistics are alarming. In 2016, the world classified one-third of its population as obese. Severe obesity, often referred to as adiposity in medical terms, is particularly concerning. It's a leading cause of diabetes, high blood pressure, and cardiovascular disease, as well as one of the most preventable causes of cancer. According to a scientific study, obesity accounts for nearly 20% of all cancer-related deaths [26–28].

As body fat increases, so does the production of hormones like estrogen in adipose tissue. Elevated estrogen levels can fuel breast cancer development, especially after menopause, when estrogen levels naturally decline. Obesity, independent of hormone dependency, is associated with a poorer prognosis for all forms of breast cancer. Inflammatory responses within adipose tissue play a crucial role in elevating cancer risk [26].

The connection between obesity and cancer is now as undeniable as the link between smoking and cancer. The risk of breast and bowel cancers is particularly high. A meta-analysis of over 82 studies found that the mortality rate among very obese breast cancer patients is 41% higher than in women of normal weight [29, 30]. Excess body fat fosters the growth of cancer cells by inducing metabolic and oxidative stress, hormonal imbalances, and chronic systemic inflammation.

Obesity also has significant health consequences for men. The conversion of testosterone to estrogen in adipose tissue lowers testosterone levels, increasing the risk of aggressive prostate tumors and leading to a poorer overall prognosis for overweight men. Although testosterone-blocking treatments are effective for prostate cancer, the reasons for the increased risk for obese men remain unclear and are the subject of ongoing research.

Additionally, severe obesity complicates early cancer detection. Overweight men tend to have lower PSA levels and an enlarged prostate, making it harder to detect prostate cancer early. The same challenge applies to breast cancer in overweight women, where larger breast sizes can obscure early signs.

Obesity significantly heightens the risk of inflammation. Fat tissue plays a substantial role in creating inflammatory reactions that can encourage tumor growth. These inflammatory processes, which are driven by metabolic products and immune cells in overloaded adipose tissue, can have an impact on the entire body.

In summary, obesity is a significant risk factor for numerous cancers. Combating obesity through a healthy diet, regular physical activity, and a mindful lifestyle is crucial not only for preventing cardiovascular disease but also for reducing cancer risk. Raising awareness about the health dangers of obesity and taking proactive steps to mitigate these risks is vital for preventing serious illness.

8.8 Ultra-Processed Food

Ultra-processed food (UPF) [31] refers to mass-produced food items that have undergone multiple physical, biological, and chemical processes before being packaged. These foods frequently come in the form of snacks, sodas, and frozen meals. They are typically vacuum-sealed and enriched with preservatives and flavor enhancers. Additionally, UPFs generally have more fat content, added sugar, salt [32], and stabilizers than their unprocessed counterparts. While these processing techniques may make the food more convenient and affordable, they also alter its physiological properties in potentially harmful ways. For example, many UPFs contain artificial sweeteners, which have been linked to diseases such as cancer. It is known that the smell or anticipation of food alone is enough to increase insulin levels in humans. Since high insulin and IGF-1 levels can shorten lifespan, it could be possible that the smell and taste of industrially processed foods, which are often enriched with chemical additives to improve smell and taste, is enough to accelerate the aging process by regularly releasing insulin [26].

Furthermore, the high fat and sugar content of UPFs can lead to obesity and other health problems. Consequently, it is crucial to know what risks come with consuming UPFs before making them a regular part of your diet.

One of the most troubling side effects of consuming processed foods is their impact on our endocrine system. Many processed foods contain high levels of additives and preservatives, which can disrupt hormones and cause various problems, including weight gain, infertility, and cancer. In addition, processed foods are often designed for immediate consumption, which can lead to overeating and increase the risk of obesity. Furthermore, these foods are typically high in sugar and unhealthy fats. They thus damage the immune system and lead to chronic diseases such as diabetes and heart disease. The time has come for the scientific community to take a closer look at the potential harm processed foods can cause and to start sharing this information with the public.

8.9   Dietary Supplements

Dietary supplements are substances that people consume in addition to their regular diet. These supplements can come from proteins, vitamins, minerals, or other substances. They are frequently consumed in pill, capsule, or powdered form. Some people need dietary supplements because they have nutrient deficiencies. Their doctor may recommend that they take specific supplements to improve their health. For example, people with anemia may need iron and folic acid supplements. In contrast, those who need to improve their bone health may need calcium and vitamin D supplements. Dietary supplements can be beneficial for some people. Still, it is crucial to speak with a doctor before taking any such supplements.

It is essential to be choosy when it comes to supplements. With all the different therapeutic claims their manufacturers make, it can be hard to decipher which supplements are backed by science and which are not. Many manufacturers claim supplements have the same health benefits as eating plant-based foods. However, it is not uncommon to see supplement users replace eating fruits and vegetables with tablets at the cost of their health. They may even assume that a single pill can replace a high-quality diet, which is false. This assumption is dangerous because it can give people a false sense of security.

In reality, many large-population studies have concluded that these supplements do not bring positive health effects. Furthermore, sometimes they cause adverse effects. And while there are plenty of legitimate reasons to take supplements, such as filling nutrient gaps in your diet, there is also a downside to this largely unregulated industry. Indeed, some products have been reported as increasing cancer risk and mortality. For example, high-dose beta-carotene supplements are linked to lung cancer [33], selenium, and vitamin E supplements are linked to prostate cancer, and a high intake of vitamin E supplements is linked to lung cancer in smokers [34]. It's crucial to prioritize a balanced diet over supplements as a source of nutrition. While taking a mix of vitamins for a short period may not cause any problems, replacing food with supplements can lead to deficiencies and other health issues. The only exception to this is when it's medically necessary to replace food with supplements. As such, it's essential to recommend supplements—notably iron, B12, and vitamin D—to vegetarians/vegans to ensure they receive all the necessary nutrients to maintain good health. Opting for supplements can prevent deficiencies from developing and promote healthy living.

A Box for Vitamin D

Although the human body can produce vitamin D itself through sunlight, foods such as nuts, fish, and mushrooms are also good sources of this important vitamin. Studies show that regular intake of vitamin D can reduce the risk of breast and bowel cancer. Studies have also documented a positive effect on bone density and immune function [35].

Recent studies from London [36, 37] and Germany indicate that vitamin D and certain gut bacteria might enhance the effectiveness of immunotherapies, particularly for skin cancer. Research in London involving mice demonstrated that removing a protein that binds vitamin D slowed tumor growth, suggesting that vitamin D could boost immune activity. Other animals, if not treated with antibiotics, could potentially inherit this effect through fecal microbiota transfer. In Germany, the DKFZ discovered that administering vitamin D3 to cancer patients reduced inflammatory markers, particularly tumor necrosis factor alpha and C-reactive protein. Furthermore, vitamin D3 has the ability to inhibit the reactivation of the Epstein-Barr virus, a pathogen associated with cancer [35].

Researchers believe that vitamin D3 supplementation might help suppress tumor-promoting inflammation. Since vitamin D deficiency is common, especially among cancer patients, regular supplementation could lower the risk of death by 12%. High levels of inflammatory markers are associated with poor disease outcomes, and vitamin D may help counteract this. However, it's important to be cautious with dosage as vitamin D is fat-soluble and can accumulate in fatty tissue and potentially cause toxic effects if taken in excess [35].

8.10 Food Synergy

It is no secret that what we eat directly impacts our health. For years, doctors and nutritionists have been advising us to eat more fruits and vegetables for their many nutritional benefits. New research indicates that there may be even more excellent benefits from consuming various fruits and vegetables. This is because of something called *food synergy*. Food synergy occurs when the different nutrients in foods work together to provide more health benefits than they would if the nutrients were consumed separately. This is because the nutrients in foods interact with each other and our bodies in ways that can boost our health. For example, one food's antioxidants may help improve the absorption of antioxidants from another. Alternatively, the phytochemicals in one nutrient may help protect the cells in our body from damage caused by the phytochemicals in another food.

High concentrations of reactive oxygen compounds and radicals have the ability to cause damage to cells and tissues, which can promote cancer development over time. In contrast, these compounds in low concentrations play an important role in the metabolism and immune system. It is advisable to consume antioxidants through a balanced diet rather than through supplements, as the latter can easily lead to an overdose.Antioxidants can be found in vegetables, fruit, whole grains, and vegetable oils and can also be taken in the form of tablets such as selenium, polyphenols, and vitamin C. Studies have shown that long-term use of antioxidants does not reduce the risk of cancer or increase life expectancy. There is even research suggesting that a high intake of antioxidants may increase cancer risk and shorten life expectancy.

There is also evidence that food synergy may help reduce cancer risk [6]. One study found that people who ate a diet rich in various fruits and vegetables had a lower risk of developing cancer than those who ate a diet lacking in variety. The researchers believe that this is because the different nutrients in fruits and vegetables work together to protect cells from damage and repair the DNA damage that can lead to cancer. So, to lower your risk of cancer, aim to meet your nutritional needs through diet alone as much as possible. Eat a wide variety of fruits and vegetables to benefit most from their nutrients. Do not forget to enjoy their delicious flavors, too!

However, it is important to recognize that nutritional recommendations are not one-size-fits-all. Depending on your situation, your dietary needs may vary. If you are in good health, maintaining a balanced diet that is suited to your physical activity, habits, preferences, and budget is crucial. However, if you have a special diet or experience digestive problems, being mindful of what you eat becomes even more important. Replacing certain foods with others to avoid nutritional deficiencies or taking supplements may be necessary. And for those at the end of life, the focus should be on pleasure as long as there are no digestive issues. In such cases, preventive measures are not necessary. Remember to make informed choices when it comes to your diet and take steps to ensure that you are meeting your nutritional needs.

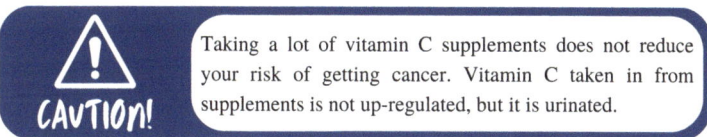

CAUTION! Taking a lot of vitamin C supplements does not reduce your risk of getting cancer. Vitamin C taken in from supplements is not up-regulated, but it is urinated.

8.11 Nutrition in the Fight Against Cancer

What Do I Need to Eat?

Anyone who wants to improve their health or heal an illness needs to answer this question first. After answering it, you should speak with your attending doctor and a nutrition specialist. The food your body and soul need and the amount of fasting (no food intake) required largely depend on your health status and if you are affected by cancer or other determinants. At least five types of health status can be presented here.

Status 1 "Not only do I feel great mentally, but physically, I am in excellent health. I have a good metabolism and no allergies that keep me from enjoying life to the fullest. Cancer runs in my family, so my goal is to avoid it at all costs!" Status 2 "Although I am generally healthy, I have difficulty with my appetite and digestion. I want to eat without experiencing nausea or stomach pain and have regular bowel movements." Status 3 "My health is not great. I have lost weight, and since my operation or chemotherapy, I have had enormous problems eating and digesting food properly. My muscles are weak, and I am sure the density of my bones has also decreased. My primary goal is to gain weight again and to enjoy food once more. I need more energy - more strength, specifically in my legs and arms." Status 4 "I am aware that my time on this earth is limited. I believe that food should provide me with happiness and not just be a means of sustenance. My goal is to live a peaceful life rather than existing simply to sustain myself until death." Status 5 "After enduring cancer and recognizing that it was partly due to my unhealthy lifestyle, I am now keen to learn all I can about living healthily. I hope to protect myself from future cancer diagnoses by learning a lot more about proper nutrition, fasting protocols and regular exercise." As discussed previously, a poor diet significantly increases cancer risks. Although we cannot always control other factors that contribute to cancer, our diets are something that we CAN control. Eating healthier foods and increasing our intake of substances found in certain food groups can help strengthen our body's resilience. It may even prevent the disease from developing or at least slow its progression. Plants and other healthy foods contain beneficial molecules that can help fight tumor progression. They do that by creating an unfavorable environment for tumors to grow.

Scientists call these beneficial molecules ***phytochemicals***, chemicals that work positively for human health. Studies indicate that phytochemicals create an environment where tumors cannot thrive. Phytochemicals do that by lowering inflammation, stopping oxidation, slowing cell reproduction and growth, and cutting off new blood vessels from supplying cancerous cells (angiogenesis). Phytochemicals are chemicals found in food and one class of phytochemicals is called ***polyphenols***.

Polyphenols can be found in many fruits, vegetables, and other foods such as berries, soybeans, turmeric root powder, dark chocolate, and cocoa products made from cacao beans and green tea made from Camellia sinensis leaves (Fig. 8.7). Adding these foods that contain polyphenols to your diet might lower your cancer risk because some substances in these foods (e.g., anthocyanins in berries) are thought to have properties protective against cancer (such as the ability to suppress the growth or division of cancer cells) [38].

As mentioned above, many types of food contain polyphenols. For instance, green tea contains a polyphenol called EGCG. Although more research must be done to determine how green tea could potentially prevent cancer, some studies have found that EGCG might help prevent cancer by suppressing inflammation, angiogenesis, and metastasis [39]. Next, the anti-inflammatory turmeric contains a polyphenol called curcumin. Curcumin lowers cyclooxygenase-2 or COX-2, which is linked to tumor growth. It also disrupts cancerous cell signaling pathways known to promote tumorigenesis [40]. But although these claims are promising, they are based on experiments conducted

Fig. 8.7 Polyphenols from food sources. Different kinds of polyphenols exist in nature and can be obtained by consuming certain types of food

on cell cultures in a laboratory setting. More human trials are thus necessary to corroborate these findings. Curcumin products and supplements have not been approved by the FDA as a cancer treatment because there is a lack of evidence that these can prevent or cure cancer. In "dietary" doses, when turmeric is added to food, it may provide a protective effect without toxicity. However, in high doses, such as when taken as a supplement or concentrate, it can be dangerous and lead to hemorrhage and hematoma.

While scientific studies may not provide conclusive evidence that consuming polyphenol-rich foods is protective against cancer, there is a growing body of research that suggests incorporating these foods into one's diet can have potential health benefits without causing harm. Unfortunately, conducting clinical studies in nutrition is challenging due to the difficulty of ensuring long-term adherence to specific diets and following a large cohort of participants over time. However, reliable scientific data in this field will likely become available in the future. While there is no guarantee that adding polyphenols to one's diet will reduce cancer risk, preclinical data on cells and animals, combined with limited but promising small-scale studies, suggest that it is worth considering as a preventive measure.

As for soybeans, the phytonutrients in soybeans are called *isoflavones*. Isoflavones are more commonly called "phytoestrogens" because they mirror the female estrogen hormone [41]. The most plentiful isoflavone in soy is genistein, which has been shown to potentially impede angiogenesis and metastasis, jumpstart signaling pathways connected with carcinogenesis [42], and affect apoptosis and the cell cycle. Genistein may prevent the growth of tumors by blocking estrogen from binding to hormone receptors on precancerous cells [43]. This contains the enzymes that trigger uncontrolled cell proliferation, resulting in reduced tumor growth. Phytoestrogens, like those found in soy, can significantly impact the growth of cancers with estrogen receptors. Depending on the amount consumed, they may block or stimulate cancer growth.

Although generally protective, these molecules can sometimes have the opposite effect. However, soy consumption in dietary form and, in moderate amounts, has been associated with reduced breast cancer rates, as observed in Asian populations. On the other hand, genistein concentrates in tablet form can stimulate the estrogen receptor and should be avoided, especially in hormone-sensitive cancer patients. Despite its potential anticancer effects, it's best to avoid taking large quantities of any substance that presents benefits and risks. A safer option is to consume smaller, dietary doses that offer fewer health risks.

Several studies have found that combining multiple isoflavones is more effective than having just one. For example, one study suggested that it is more beneficial to eat whole foods that contain a combination of active isoflavone compounds rather than eat individual compounds found in supplements [44]. Although some studies have found that soy and genistein may help prevent cancer, other studies disagree [45, 46]. Sometimes, it is challenging to know how trustworthy the findings of epidemiological studies are because they often rely on people self-reporting their dietary habits. However, that method is not always reliable, and such studies do not consider genetic or behavioral variables. Therefore, we need more research to understand better how effective genistein from soy is at preventing cancer and what a safe dose would be.

Plant-based foods taste great, but they are also full of *bioactive phytochemicals* that can help reduce the effects of carcinogens on our bodies. These phytochemicals are anticarcinogenic, and they are found in garlic, citrus fruits, and cruciferous vegetables. When we add these foods to our recipes, we are making delicious and healthy meals. Studies have shown that consuming cruciferous vegetables forms a compound called sulforaphane, which is known to be protective against cancer-causing substances [47]. In addition, the bio-active compounds in these vegetables are anti-inflammatory.

Although human studies on this matter have mixed results, there is still some evidence to support the benefits of cruciferous veggies [48]. Indeed, consuming Omega-3-rich foods may help reduce the risk of chronic diseases such as cancer because those foods are a good source of healthy essential fats called omega-3. These are essential fats that can be sourced from flaxseed, perilla, and canola oils. Fatty fish such as sardines, mackerel, and salmon are also good sources of Omega-3. In addition, omega-3 may also possess pro-apoptotic, antiproliferative, and antiandrogenic properties. Finally, omega-3 is a valuable energy supplier and a potent anti-inflammatory [49]. Since inflammation is an essential factor that promotes cancer, anti-inflammatories are important in fighting cancer. However, there is no convincing data to prove a clear benefit in humans. Recommendations about the value of omega-3 supplements, found in certain vegetable oils and fatty fish, for cancer prevention, are not based on solid scientific evidence [50].

While we cannot wholeheartedly recommend any one dietary change or food to protect against cancer, nutrition-based cancer prevention has become a growing field of scientific research in recent decades. Researchers have generated mountains of evidence showing that diet can prevent mutated cells from becoming metastatic. They have even identified several bioactive compounds, which interfere with cancer development. In the majority of their experiments, they use animals and cells grown in the laboratory because studies on human subjects are not as reliable due to many variables. Even so, the findings are worth considering. We would all like to believe that there is a food whose consumption prevents cancer. However, based on the scientific literature, it's impossible to unequivocally recommend a dietary change, a specific food, or a particular component to protect against cancer. Several nutrients seem promising based on studies conducted on cells or animals, although they have not yet definitively proven their efficacy in humans. Including these nutrients in our diet in their natural, unprocessed form, and in moderation, is certainly advisable, as long as consuming them in "dietary" doses isn't harmful. However, it's important to maintain a varied diet to prevent deficiencies. Nevertheless, cancer prevention through nutrition has become a rapidly growing area of scientific research in recent decades, and new data will likely emerge in the future.

8.12 Gut Microbiome

The human microbiome is a complex ecosystem of almost 40 trillion micro-organisms, consisting of around 3000 species [51]. It plays a vital role in bodily functions, such as digestion, nutrient absorption, and drug efficacy. The bacterial component of the human microbiome alone is estimated to exceed the number of cells in the body. The dynamic composition of the microbiome, influenced by factors such as age, geographic location, and diet, underscores its importance in maintaining systemic balance and functional stability in the human body [52].

Our eating habits significantly impact our microbiome's health. Factors such as diet, antibiotic treatments, invasive pathogens, drug use, and stress can disrupt the delicate balance of organisms in our body and cause dysbiosis, which is closely linked to numerous diseases, including cancer risk [53].

Current studies focus on exploring the complexity of the gut microbiome [54]. One critical finding is the link between microbial populations and increased cancer risk. For example, the presence of *Helicobacter pylori* in gastric cancer is a well-documented example of a microbial infection associated with cancer development in the human gastrointestinal tract [55]. Recent research has shown that certain bacteria, such as *Streptococcus bovis* and *Bacteroides fragilis*, may be involved in the development of colorectal cancer [56].

Maintaining a balanced and diverse gut microbiome is essential for reducing the risk of developing diseases. Including fiber-rich foods in our daily meals, such as avocado, bananas, apples, broccoli, beans, raspberries, and strawberries, provides the necessary nutrients for a thriving microbial community. Fermented foods with live microbes, such as yogurt, kefir, fermented cottage cheese, sauerkraut, kombucha tea, kimchi, and other fermented vegetables, provided the patient can digest them, of course, can also significantly improve the diversity and health of our microbiome. However, caution is advised when using commercially available probiotics, as they may not effectively colonize our gut.

Research on the microbiome in cancer treatment and prevention is promising, but further studies are needed to confirm safety and efficacy.

Now, here are four studies that reveal new findings in microbiome research. A study conducted in Berlin [54] over 2 years has highlighted the significant connection between the microbiome and cancer tests. This research involved a comprehensive analysis of the intestinal microorganisms of over 500 cancer patients. It emphasized the importance of considering a patient's microbiome in immunotherapy. Notably, the study found that cancer patients who have previously taken antibiotics may respond less effectively to treatment, a finding that has significant implications for clinical practice. A study in Munich [57] examined the microbiome signatures and diet of individuals with metastatic skin cancer undergoing ICI therapy. It found that specific microbiomes had a better response, and the side effects of immunotherapies were lower when sufficient fiber and omega-3 fatty acids were supplied.

In Hamburg [58], a groundbreaking study revealed that the effect of chemotherapy on pancreatic cancer is influenced by metabolic products in the gut, which are, in turn, influenced by our diet and microbiome. The study found that the amino acid tryptophan, commonly found in foods like turkey and eggs, correlates with the effectiveness of the therapy. This finding could potentially lead to personalized dietary recommendations for cancer patients undergoing chemotherapy, thereby improving treatment outcomes.

Australian [59] researchers discovered that bacteria such as *Escherichia coli* survive longer in tumors. They genetically modified the bacteria to produce salicylates, which enter the bloodstream via the cancer. Therapeutic possibilities are also seen in the coupling of *Escherichia coli* with nanobodies to neutralize immune checkpoints.

Despite promising results, large-scale studies are needed to determine the safety and efficacy of probiotics in cancer treatment. While some studies have suggested that certain strains of probiotics may enhance the effectiveness of cancer treatments, others have raised concerns about the potential for probiotics to interfere with the immune system's response to cancer. Therefore, the role of probiotics in cancer treatment is still unclear, which is why randomized trials are required to provide more definitive answers.

Considering the wide variety of foods available to us, it's important to recognize the profound influence of our dietary choices on our health. While a range of foods can strengthen our health and assist in preventing and fighting cancer, it is concerning to note the prevalence of convenience and processed foods laden with harmful chemicals. Overconsumption of these products can increase the risk of developing cancer. This emphasizes the importance of being mindful of our consumption of some of our favorite foods, such as alcohol, red meat, processed meats, excessive amounts of sugar, and salt, as they may be classified as carcinogenic, according to the International Agency for Research on Cancer (IARC).

Recent research has illuminated the potential involvement of specific bacteria, such as *Streptococcus bovis*, *Bacteroides fragilis*, and *Peptostreptococcus anaerobius*, in promoting colorectal cancer. These bacteria are thought to contribute to carcinogenesis through various mechanisms, including the activation of Th17 cells [60], direct DNA damage [61], and stimulation of cholesterol synthesis [62]. Additionally, the microbiota, including *Methylobacterium radiotolerans*, is speculated to play a role in breast carcinogenesis, potentially driving the condition through their metabolites [63].

In the chapter titled "Nutrition in the Fight against Cancer," we aim to direct attention to two specific health conditions of cancer patients that hold crucial significance in the realm of nutrition. These conditions, denoted as status three and status four, necessitate a tailored nutritional approach to enhance the quality of life of patients and address their evolving needs throughout the course of their illness.

The human microbiome is a complex ecosystem of nearly 40 trillion microorganisms and around 3000 species. It plays an active role in various bodily functions such as digestion, nutrient absorption, and medication effectiveness. It has been estimated that the bacterial component of the human microbiome alone exceeds the number of human cells in the body. Its dynamic composition, which is influenced by age, geographical location, and diet, underscores its importance in maintaining systemic balance and functional stability within the human body.

Our dietary choices not only power our bodies but also play a crucial role in the health of our microbiome. On the other hand, factors like diet, antibiotic treatments, invasive pathogens, drug use, or stress can upset the delicate balance of organisms within our bodies, causing dysbiosis. These disturbances have a profound impact on the intestinal microbiome and are strongly associated with numerous diseases, including the potential risk of cancer.

It's crucial to prioritize a balanced and diverse gut microbiome to lower the risk of developing diseases. Our diet significantly influences the composition and function of our gut microbiome. Including a wide variety of fiber-rich foods in our daily meals, such as avocado, banana, apple, broccoli, beans, raspberries, and strawberries, is essential to provide the necessary nutrients for a thriving microbial community. Fiber undergoes fermentation in the colon, producing short-chain fatty acids with potent anti-inflammatory properties. It's highly recommended to incorporate at least thirty different fiber-rich foods into our weekly diets, including various vegetables, fruits, nuts, herbs, and seeds. By diversifying our nutrition, we can create an environment where harmful microbes struggle to dominate, ultimately reducing the risk of disease. Furthermore, integrating fermented foods containing live microbes into our diet can significantly enhance the diversity and health of our microbiome. However, it's crucial to exercise caution, as many store-bought probiotics may not effectively colonize our guts.

Nutrition in Particular Situations for Cancer Patients

Unfortunately, just as various foods can improve our health and help us face cancer, there are also many convenience foods we have grown to love, which are loaded with harmful chemicals. Consuming them can lead us to develop cancer. Indeed, some of our favorite items, such as alcohol, meat, and sausages, or sugar and salt, must be consumed moderately. This is because they are carcinogenic, according to the International Agency for Research on Cancer (IARC). That said, when we began the section in this chapter called "Nutrition in the Fight Against Cancer," we listed five types of health status cancer patients might have. We will discuss two of them in more depth here.

Status 3

As you may recall, this is how we described health status 3 earlier in this chapter:

> My health is not great. I have lost weight, and since my operation or chemotherapy, I have had enormous problems eating and digesting food properly. My muscles are weak, and I am sure the density of my bones has also decreased. My primary goal is to gain weight again and to enjoy food once more. I need more energy - more strength, specifically in my legs and arms.

Dining is a social experience for many people and can positively impact well-being. It is an essential part of social dynamics and enhances the quality of life. However, illness or treatment may make it difficult for patients to eat and drink as usual. This can lead to isolation from loved ones, misunderstandings, and conflict. If possible, try not to let the patient eat alone. Eating with others usually improves appetite, too. This is important for patients with health status 3 who want to gain weight and thus gain strength. Also, it is good to create a relaxed atmosphere during meal times whenever possible.

While undergoing cancer treatment, patients should steer clear of highly processed foods that contain hydrogenated oils. These oils have been known to exacerbate inflammation. Furthermore, those fighting cancer may have a weakened immune system for days or even years after treatment. As such, they should avoid eating raw proteins, which carry the risk of bacteria and food poisoning. In addition to practicing good hygiene, additional precautions should be taken when storing and preparing food and also dining out. This helps to avoid contamination and protect immunocompromised patients from contracting infectious diseases. Diseases could harm them and sometimes delay the treatment.

Status 4

As you may recall, this is how we described health status 4 earlier in this chapter:

> I am aware that my time on this earth is limited. I believe that food should provide me with happiness and not just be a means of sustenance. My goal is to live a peaceful life rather than existing simply to sustain myself until death.

When a patient's appetite shifts, trying to make them eat might be stressful. Instead, plan meals around when they feel the most comfortable and can eat more efficiently. Additionally, many cancer patients reduce their intake because food preparation is too tiring. If someone else offers to do the grocery shopping or cooking, it could be a big help. Suppose the disease is incurable and has progressed to a late stage. The patient could thus have health status 4. In that case, palliative care's primary goal is to mask symptoms and reduce suffering [64]. By doing so, we improve not just the patient's quality of life but also the lives of those around them.

Mealtimes can be a challenge for patients and their loved ones, especially when emotions come into play, which is often the case. However, the situation can be improved by addressing common concerns, such as the fear of not eating enough, the fear of overeating, the fear of being rejected for not eating everything that's offered, or the desire for something different than what's been prepared. Talking with medical staff and loved ones can help alleviate these worries and create a more relaxed atmosphere during meals.

In more severe cases, it is necessary to give the patient adequate nutrition for them to have a chance to live longer. Even though this is done, their physical condition will still get worse as time goes on. The patients might not have any appetite or desire for food or drink. Instead of trying to force nourishment, we provide "comfort feeding." This means giving them foods they can handle easily or their favorite dishes. Doing so can improve their quality of life and take care of some symptoms.

Knowing that those affected by this will not die of hunger or thirst can be reassuring. In most cases, they need some fluids and good oral hygiene. Keeping their mucous membranes and lips moist is also helpful. In summary, nutritional care during the end-of-life process should be based on the patient's needs and preferences to improve their quality of life through comfort.

Eating the right foods is crucial for cancer patients before, during, and after treatment. The type of treatment received will determine what dietary advice should be followed. For example, if someone has mouth ulcers, they should avoid spicy and acidic foods instead of cold foods. If bloating is an issue, avoiding foods that can cause flatulence, like cabbage, peas, beans, and lentils, is best. For those receiving chemotherapy, it's essential to be cautious to prevent digestive infections by eating fresh foods, cooking meat and fish thoroughly, and avoiding shellfish. Maintaining proper hygiene while preparing food and paying attention to expiration dates can also help reduce the risk of digestive diseases. Studies indicate that a diverse diet, including fruits and vegetables, is particularly beneficial for those undergoing immunotherapy. Fermented foods can also be included in the diet from time to time. Overall, it's recommended that cancer patients consume protein-rich foods such as meat, fish, eggs, tofu, and protein supplements. This helps the body to repair and defend itself during cancer treatment. It's important to note that dietary recommendations are personalized and discussed with healthcare professionals.

8.13 Okinawa Excursion

In addition to helping reduce obesity and lowering insulin and inflammation levels, a low-calorie diet may also help prevent cancer by altering hormone balance or the gut flora. Studies have linked food scarcity to a lower risk of colorectal and breast cancer in human populations. However, finding volunteers for strictly controlled, long-term diets can be difficult. As a result, scientists often rely on analyzing epidemiological data from people who lived through periods of war, famine, or restricted food intake.

The people of Okinawa, part of Japan's Centennial Islands, are well-known for their high-quality, low-calorie diet [65]. Okinawans primarily consume fresh vegetables, rice, seafood, seaweed, and tea. Women in Okinawa have a significantly lower risk of breast cancer compared to those in other parts of Japan, are largely free from modern "diseases of civilization," and maintain their mobility well into old age [66]. Researchers often find mutations in genes related to AMP, mTOR, sirtuins, kinase, insulin, and IGF-1 in elderly individuals, such as those in Okinawa. These genes play a crucial role in regulating nutrients like sugar and growth hormones. To fully understand how these signaling pathways influence lifespan and whether medications could replicate or even enhance the benefits of a calorie-restricted diet, more research is necessary.

An Okinawa-inspired diet, particularly the Japanese approach, could potentially replace the Mediterranean diet as the gold standard for healthy eating, gaining popularity among younger generations. Policymakers could encourage healthier eating habits among young people, steering them away from fast food and toward more nutritious choices. Another approach would be to identify compounds in the diets of long-lived populations that may have cancer-preventive properties.

Summary: This chapter underscores the importance of nutrition in health, particularly in the fight against cancer. Nutrition isn't just about survival and pleasure; it's a key player in disease prevention. The quality of water, air, and food are all factors, and a diet rich in fruits and vegetables can significantly reduce the risk of cancer. However, it's important to note that processed meat and alcohol can increase the risk. The key here is that a balanced diet, along with the *Hara Hachi Bu* concept, can have positive effects. Foods like sugar, salt, and ultra-processed products should be consumed in moderation. Remember, food supplements should only be taken under medical advice. Polyphenols, phytochemicals, and the gut microbiome may all play a role in cancer prevention and treatment. For cancer patients, an adapted diet can be a game-changer, helping to alleviate symptoms and improve quality of life.

References

1. Pang Y, et al. Associations of adiposity and weight change with recurrence and survival in breast cancer patients: a systematic review and meta-analysis. Breast Cancer. 2022;29(4):575–88. https://doi.org/10.1007/s12282-022-01355-z.
2. Eating fast increases diabetes risk. ScienceDaily. https://www.sciencedaily.com/releases/2012/05/120507210038.htm. Accessed 05 July 2024.
3. Srivastava SP, Goodwin JE. Cancer biology and prevention in diabetes. Cells. 2020;9(6):1380. https://doi.org/10.3390/cells9061380.
4. Alcohol and cancer risk fact sheet—NCI. https://www.cancer.gov/about-cancer/causes-prevention/risk/alcohol/alcohol-fact-sheet. Accessed 05 July 2024.
5. Don't eat until you're full—instead, mind your hara hachi bu point. Cleveland Clinic. https://health.clevelandclinic.org/dont-eat-until-youre-full-instead-mind-your-hara-hachi-bu-point. Accessed 05 July 2024
6. Summary-of-Third-Expert-Report-2018.Pdf.
7. Allen NE, et al. Moderate alcohol intake and cancer incidence in women. J Natl Cancer Inst. 2009;101(5):296–305. https://doi.org/10.1093/jnci/djn514.
8. Bagnardi V, et al. Light alcohol drinking and cancer: a meta-analysis. Ann Oncol. 2013;24(2):301–8. https://doi.org/10.1093/annonc/mds337.
9. Weinberg RA. The biology of cancer. 2nd ed. New York: W.W. Norton & Company; 2013. https://doi.org/10.1201/9780429258794.
10. Britt KL, Cuzick J, Phillips K-A. Key steps for effective breast cancer prevention. Nat Rev Cancer. 2020;20(8):417–36. https://doi.org/10.1038/s41568-020-0266-x.

11. Avis d'experts relatif à l'évolution du discours public en matière de consomma-tion d'alcool en France organisé par Santé publique France et l'Institut national du cancer. https://www.santepubliquefrance.fr/liste-des-actualites/avis-d-experts-relatif-a-l-evolution-du-discours-public-en-matiere-de-consommation-d-alcool-en-france-organise-par-sante-publique-france-et-l-insti. Accessed 10 July 2024.
12. Bouvard V, et al. Carcinogenicity of consumption of red and processed meat. Lancet Oncol. 2015;16(16):1599–600. https://doi.org/10.1016/S1470-2045(15)00444-1.
13. Patrick K. How does processed meat cause cancer and how much matters? Cancer Research UK—Cancer News. https://news.cancerresearchuk.org/2021/03/17/bacon-salami-and-sausages-how-does-processed-meat-cause-cancer-and-how-much-matters/. Accessed 05 July 2024.
14. Cross AJ, Sinha R. Meat-related mutagens/carcinogens in the etiology of colorec-tal cancer. Env Mol Mutagen. 2004;44(1):44–55. https://doi.org/10.1002/em.20030.
15. Chemicals in meat cooked at high temperatures and cancer risk—NCI. https://www.cancer.gov/about-cancer/causes-prevention/risk/diet/cooked-meats-fact-sheet. Accessed 05 July 2024.
16. Nitrosamines in food raise a health concern | EFSA. https://www.efsa.europa.eu/en/news/nitrosamines-food-raise-health-concern. Accessed 04 July 2024.
17. IARC. Red meat and processed meat.
18. Tasevska N, et al. Sugars in diet and risk of cancer in the NIH-AARP diet and health study. Int J Cancer. 2012;130(1):159–69. https://doi.org/10.1002/ijc.25990.
19. Giovannucci E, et al. Diabetes and cancer: a consensus report. CA Cancer J Clin. 2010;60(4):207–21. https://doi.org/10.3322/caac.20078.
20. Vigneri P, et al. Diabetes and cancer. Endocr Relat Cancer. 2009;16(4):1103–23. https://doi.org/10.1677/ERC-09-0087.
21. Hu J, et al. Glycemic index, glycemic load and cancer risk. Ann Oncol. 2013;24(1):245–51. https://doi.org/10.1093/annonc/mds235.
22. Giovannucci E. Insulin and colon cancer. Cancer Causes Control. 1995;6(2):164–79. https://doi.org/10.1007/bf00052777.
23. Salt in your diet. nhs.uk. https://www.nhs.uk/live-well/eat-well/food-types/salt-in-your-diet/. Accessed 06 July 2024.
24. Nutrition, C. for F. S. and A. Sodium in your diet. FDA 2024.
25. Strnad M. Salt and cancer. Acta Medica Croat Časopis Hrvatske Akad Med Znan. 2010;64:159–61.
26. Heikenwälder H, Heikenwälder M. Der moderne Krebs—Lifestyle und Umweltfaktoren als Risiko. Berlin/Heidelberg: Springer Berlin Heidelberg; 2023. https://doi.org/10.1007/978-3-662-66576-3.
27. Aggarwal BB, Vijayalekshmi RV, Sung B. Targeting inflammatory pathways for prevention and therapy of cancer: short-term friend, long-term foe. Clin Cancer

Res Off J Am Assoc Cancer Res. 2009;15(2):425–30. https://doi.org/10.1158/1078-0432.CCR-08-0149.

28. Calle EE, Rodriguez C, Walker-Thurmond K, Thun MJ. Overweight, obesity, and mortality from cancer in a prospectively studied cohort of U.S. adults. N Engl J Med. 2003;348(17):1625–38. https://doi.org/10.1056/NEJMoa021423.

29. Kohls M, Freisling H, Charvat H, et al. Impact of cumulative BMI and cardio-metabolic diseases on survival in colorectal and breast cancer: a multi-centre cohort study. BMC Cancer. 2022;22(1):546. https://doi.org/10.1186/s12885-022-09589-y.

30. Chan, et al. Body mass index and survival in women with breast cancer: systematic review and meta-analysis. Ann Oncol. 2014;25(10):1901–14.

31. Lane MM, et al. Ultra-processed food exposure and adverse health outcomes: umbrella review of epidemiological meta-analyses. BMJ. 2024:384. https://doi.org/10.1136/bmj-2023-07731.

32. Fiolet T, et al. Consumption of ultra-processed foods and cancer risk: results from NutriNet-Santé prospective cohort. BMJ. 2018:k322. https://doi.org/10.1136/bmj.k322.

33. Do not use supplements for cancer prevention | Cancer Prevention | WCRF International. https://www.wcrf.org/diet-activity-and-cancer/cancer-prevention-recommendations/do-not-use-supplements-for-cancer-prevention/. Accessed 05 July 2024.

34. Vitamins, diet supplements and cancer. https://www.cancerresearchuk.org/about-cancer/treatment/complementary-alternative-therapies/individual-therapies/vitamins-diet-supplements. Accessed 05 July 2024.

35. zur Hausen H, Bund T, de Villiers E-M. Infectious agents in bovine red meat and Milk and their potential role in cancer and other chronic diseases. In: Hunter E, Bister K, editors. Viruses, genes, and cancer. Cham: Springer International Publishing; 2017. p. 83–116. https://doi.org/10.1007/82_2017_3.

36. Gwenzi T, Zhu A, Schrotz-King P, et al. Effects of vitamin D supplementation on inflammatory response in patients with cancer and precancerous lesions: systematic review and meta-analysis of randomized trials. Clin Nutr. 2023;42(7):1142–50. https://doi.org/10.1016/j.clnu.2023.05.009.

37. Giampazolias E, et al. Vitamin D regulates microbiome-dependent cancer immunity. Science. 2024;384(6694):428–37. https://doi.org/10.1126/science.adh7954.

38. Skrovankova S, et al. Bioactive compounds and antioxidant activity in different types of berries. Int J Mol Sci. 2015;16(10):24673–706. https://doi.org/10.3390/ijms161024673.

39. Singh BN, et al. Green tea catechin, epigallocatechin-3-gallate (EGCG): mechanisms, perspectives and clinical applications. Biochem Pharmacol. 2011;82(12):1807–21. https://doi.org/10.1016/j.bcp.2011.07.093.

40. Zoi V, et al. The role of curcumin in cancer treatment. Biomedicines. 2021;9(9):1086. https://doi.org/10.3390/biomedicines9091086.

41. Křížová L, et al. Isoflavones Molecules. 2019;24(6):1076. https://doi.org/10.3390/molecules24061076.
42. Sarkar FH, Li Y. Soy isoflavones and cancer prevention. Cancer Investig. 2003;21(5):744–57. https://doi.org/10.1081/cnv-120023773.
43. Desmawati D, Sulastri D. Phytoestrogens and their health effect. Open Access Maced J Med Sci. 2019;7(3):495–9. https://doi.org/10.3889/oamjms.2019.044.
44. Hsu A, et al. Differential effects of whole soy extract and soy isoflavones on apoptosis in prostate cancer cells. Exp Biol Med Maywood. 2010;235(1):90–7. https://doi.org/10.1258/ebm.2009.009128.
45. Nakamura H, et al. Genistein increases epidermal growth factor receptor Signaling and promotes tumor progression in advanced human prostate cancer. PLoS One. 2011;6(5):e20034. https://doi.org/10.1371/journal.pone.0020034.
46. Wang J, et al. Genistein alters growth factor signaling in transgenic prostate model (TRAMP). Mol Cell Endocrinol. 2004;219(1):171–80. https://doi.org/10.1016/j.mce.2003.12.018.
47. Lenzi M, et al. Sulforaphane as a promising molecule for fighting cancer. In: Advances in nutrition and cancer; cancer treatment and research. Berlin/Heidelberg: Springer Berlin Heidelberg; 2014. p. 207–23. https://doi.org/10.1007/978-3-642-38007-5_12.
48. Cruciferous vegetables and cancer prevention—NCI. https://www.cancer.gov/about-cancer/causes-prevention/risk/diet/cruciferous-vegetables-fact-sheet. Accessed 05 July 2024.
49. Chan JK, et al. Effect of dietary A-linolenic acid and its ratio to linoleic acid on platelet and plasma fatty acids and thrombogenesis. Lipids. 1993;28(9):811–7. https://doi.org/10.1007/bf02536235.
50. Hanson, et al. Omega-3, omega-6, and total dietary polyunsaturated fat on cancer incidence: systematic review and meta-analysis of randomised trials. Br J Cancer. 2020;122(8):1260–70. https://doi.org/10.1038/s41416-020-0761-6.
51. Zhao L-Y, Mei J-X, Yu G, Lei L, Zhang W-H, Liu K, Chen X-L, Kołat D, Yang K, Hu J-K. Role of the gut microbiota in anticancer therapy: from molecular mechanisms to clinical applications. Signal Transduct Target Ther. 2023;8(1):201. https://doi.org/10.1038/s41392-023-01406-7.
52. Kandalai S, Li H, Zhang N, Peng H, Zheng Q. The human microbiome and cancer: a diagnostic and therapeutic perspective. Cancer Biol Ther. 2023;24(1):2240084. https://doi.org/10.1080/15384047.2023.2240084.
53. Ağagündüz D, Cocozza E, Cemali Ö, Bayazıt AD, Nanì MF, Cerqua I, Morgillo F, Saygılı SK, Berni Canani R, Amero P, Capasso R. Understanding the role of the gut microbiome in gastrointestinal cancer: a review. Front Pharmacol. 2023;14:1130562. https://doi.org/10.3389/fphar.2023.1130562.
54. Eng L, Sutradhar R, Niu Y, Liu N, Liu Y, Kaliwal Y, Powis ML, Liu G, Peppercorn JM, Bedard PL, Krzyzanowska MK. Impact of antibiotic exposure before immune checkpoint inhibitor treatment on overall survival in older adults with cancer: a

population-based study. J Clin Oncol. 2023;41(17):3122–34. https://doi.org/10.1200/JCO.22.00074.

55. IARC Working Group on the Evaluation of Carcinogenic Risks to Humans. Schistosomes, liver flukes and Helicobacter pylori. IARC Monogr Eval Carcinog Risks Hum. 1994;61:1–241.

56. Gagnière J, Raisch J, Veziant J, Barnich N, Bonnet R, Buc E, Bringer M-A, Pezet D, Bonnet M. Gut microbiota imbalance and colorectal cancer. World J Gastroenterol. 2016;22(2):501–18. https://doi.org/10.3748/wjg.v22.i2.501.

57. Simpson RC, et al. Diet-driven microbial ecology underpins associations between cancer immunotherapy outcomes and the gut microbiome. Nat Med. 2022;28(11):2344–52. https://doi.org/10.1038/s41591-022-01965-2.

58. Tintelnot J, et al. Microbiota-derived 3-IAA influences chemotherapy efficacy in pancreatic cancer. Nature. 2023;615(7950):168–74. https://doi.org/10.1038/s41586-023-05728-y.

59. Gurbatri CR, et al. Engineering tumor-colonizing E. Coli Nissle 1917 for detection and treatment of colorectal neoplasia. Nat Commun. 2024;15(1):646. https://doi.org/10.1038/s41467-024-44776-4.

60. Wu S, et al. A human colonic commensal promotes colon tumorigenesis via activation of T helper type 17 T cell responses. Nat Med. 2009;15(9):1016–22. https://doi.org/10.1038/nm.2015.

61. Cuevas-Ramos G, et al. Escherichia Coli induces DNA damage in vivo and triggers genomic instability in mammalian cells. Proc Natl Acad Sci USA. 2010;107(25):11537–42. https://doi.org/10.1073/pnas.1001261107.

62. Tsoi H, et al. Peptostreptococcus Anaerobius induces intracellular cholesterol biosynthesis in colon cells, promoting proliferation and dysplasia in mice. Gastroenterology. 2017;152(6):1419–1433.e5. https://doi.org/10.1053/j.gastro.2017.01.009.

63. Xuan C, et al. Microbial dysbiosis is associated with human breast cancer. PLoS One. 2014;9(1):e83744. https://doi.org/10.1371/journal.pone.0083744.

64. Cotogni P, Stragliotto S, Ossola M, Collo A, Riso S. On behalf of the intersociety Italian Working Group for nutritional support in cancer. The role of nutritional support for cancer patients in palliative care. Nutrients. 2021;13(2):306. https://doi.org/10.3390/nu13020306.

65. Kriti K. Ikigai: the Japanese secret to a long and happy life—#bookthoughts. Armed with A Book. https://armedwithabook.com/ikigai-the-japanese-secret-to-a-long-and-happy-life/. Accessed 08 Aug 2024.

66. Umar A, Dunn BK, Greenwald P. Future directions in cancer prevention. Nat Rev Cancer. 2012;12(12):835–48. https://doi.org/10.1038/nrc3397.

9

Physical Activity and Cancer

Contents

Abstract In this chapter, we will take an in-depth look at the role of physical activity in cancer prevention and treatment. We will explore how an active lifestyle can reduce the risk of certain types of cancer and how exercise can influence cancer cells' development, proliferation and survival. In addition, we will discuss the different types of exercise, such as aerobic exercise and resistance training, and their specific effects on cancer cells. We will also emphasize the supportive role of exercise in reducing side effects of cancer treatment, improving physical performance, and promoting a positive outlook on life in cancer patients. Furthermore, we will discuss the recommendations for physical activity in cancer survivors and the importance of moderate exercise for health and well-being. Finally, we will summarize the benefits of regular exercise for cancer prevention and treatment and present strategies to increase physical activity in everyday life.

9.1   Benefits of Physical Activity

As discussed in Chap. 2, "How Cancer Develops," lifestyle is the greatest factor contributing to cancer risk. In recent decades, physical activities have become a choice rather than a necessity. Consuming convenience products and having little physical activity are unfortunate signs of our entry into an age of significantly increased cancer risk. According to the World Cancer Research Fund, 20–25% of all cancers are caused by obesity and insufficient physical activity. Fortunately, such factors can be altered by changing our behavior [1, 2] and by increasing our physical activity. These are must-haves in cancer prevention and intervention plans.

Indeed, studies have consistently shown that if we have a physically active lifestyle, it protects us from developing certain cancers [3, 4]. Recent research indicates that physical activity can decrease the proliferation, aggressiveness, and survival of cancer cells in active individuals. This happens by decreasing the function of specific genes that are responsible for how cancer develops [5]. Additionally, randomized and controlled clinical trials suggest that exercise may be a valuable tool to prevent cancer and to promote longer survival times for those suffering from it [6].

Exercise is an important factor in a healthy lifestyle, and many people benefit from increased physical activity in their everyday lives for reasons unrelated to cancer. Regular physical exercise is an effective and powerful tool that can prevent and treat cancer, with its efficacy sometimes comparable to that of drug treatments. Research shows that after breast cancer treatment, regular exercise can significantly reduce the risk of recurrence (up to 35%) and mortality (up to 40%) [7, 8]. While exercise is not a cure for cancer, it can help reduce the risk of recurrence and should be strongly encouraged. Let's take control of our health and make physical activity an integral part of our daily routine! Yet, two main types of exercise have varying effects on cancer. First, vigorous aerobic exercise, also known as endurance or cardio exercise, involves the lungs, muscles, heart, and blood vessels. Studies have determined that this form of exercise helps protect us against aspects of cancer, such as its development, mortality rate, and recurrence of aggressive forms [9]. Second,

Fig. 9.1 Decreased cancer risk due to physical activity. Having a physically active lifestyle is important to protect oneself against cancer development

resistance exercise helps improve body composition by increasing muscle mass and strength, and it has even been linked to decreased metastasis [10].

Recent studies have sought to answer the pressing question of how regular exercise slows down cancer (Fig. 9.1). The evidence is not concrete, but researchers suggest that exercising may modify several components, which can lower our risk of cancer. For example, physical activity changes how muscles respond to the influence of cancer cells. These changes induce numerous adaptations in the skeletal muscles, which have protective effects on metabolism and infection. Physical activity also alters the molecular and cellular composition of blood.

Exercise can be an effective tool in the fight against cancer for many reasons [11]. First, it helps to increase adrenaline production, an important hormone that is essential for the body. When adrenaline binds to certain receptors on cancer cells, it helps control inflammation and blood vessel formation, which in turn can slow tissue invasion and tumor growth [12, 13]. In addition, adrenaline has a positive effect on immunity as it increases the killing power of natural killer cells [14]. Adrenaline also mobilizes other immune cells to move towards the tumor [15].

Exercising releases a range of hormones called ***myokines***, which are responsible for tissue repair and maintenance. One key myokine is ***decorin***, which

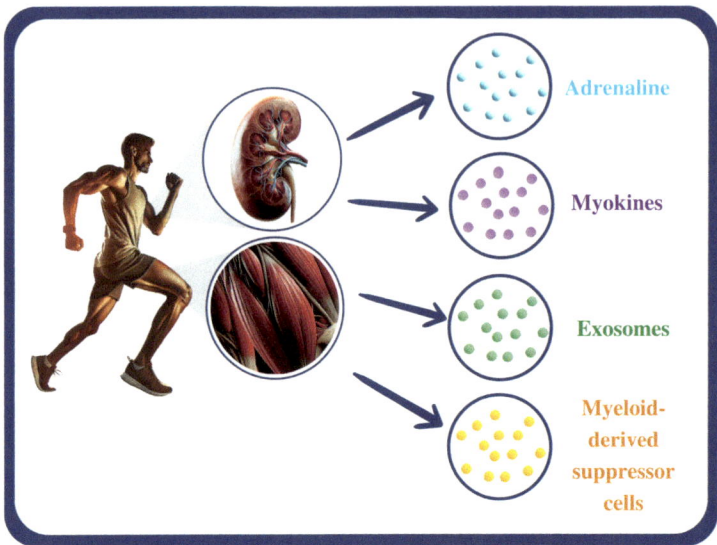

Fig. 9.2 Exercise and cancer-fighting molecules. Physical exercise causes the release of several molecules that have cancer-fighting abilities

has been found to have some impressive cancer-fighting qualities. Studies indicate that decorin exhibits anti-inflammatory, antioxidant, and antiangiogenic properties, as it can directly inhibit tumor growth [16–19] (Fig. 9.2).

Third, during physical activity, our muscles require energy to function properly. Like all cells in the body, muscles "burn" sugar (carbohydrates) in the presence of oxygen to produce energy. However, when there is not enough oxygen in the blood, our body uses an alternative mechanism to generate energy from sugar. This process produces a waste product called lactate, which is usually eliminated by the liver and kidneys. Interestingly, recent studies have shown that our body can recycle lactate even when there is enough oxygen present. However, cancer cells have devised ways to use lactate even when they have ample oxygen. Scientists believe that excess lactate in the body can help cancer form and progress via angiogenesis, immune escape, cell migration, and metastasis. Regular intervals between physical activity can help us metabolize the excess lactate in our body, thus neutralizing an essential precursor for cancer development.

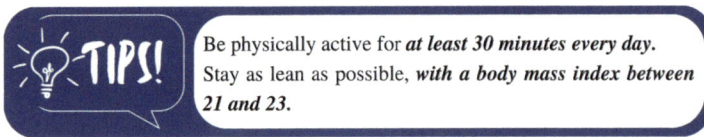

Fourth, individuals with cancer often have elevated insulin levels in their blood. Regular exercise helps reduce insulin concentrations in the bloodstream, thus decreasing the risk of developing cancer [20]. Additionally, exercise releases anticancer chemicals into our bodies that help stimulate DNA repair and fight off abnormal cells. This combination of actions works together to prevent cancer cells from multiplying and spreading further within our bodies (Fig. 9.3).

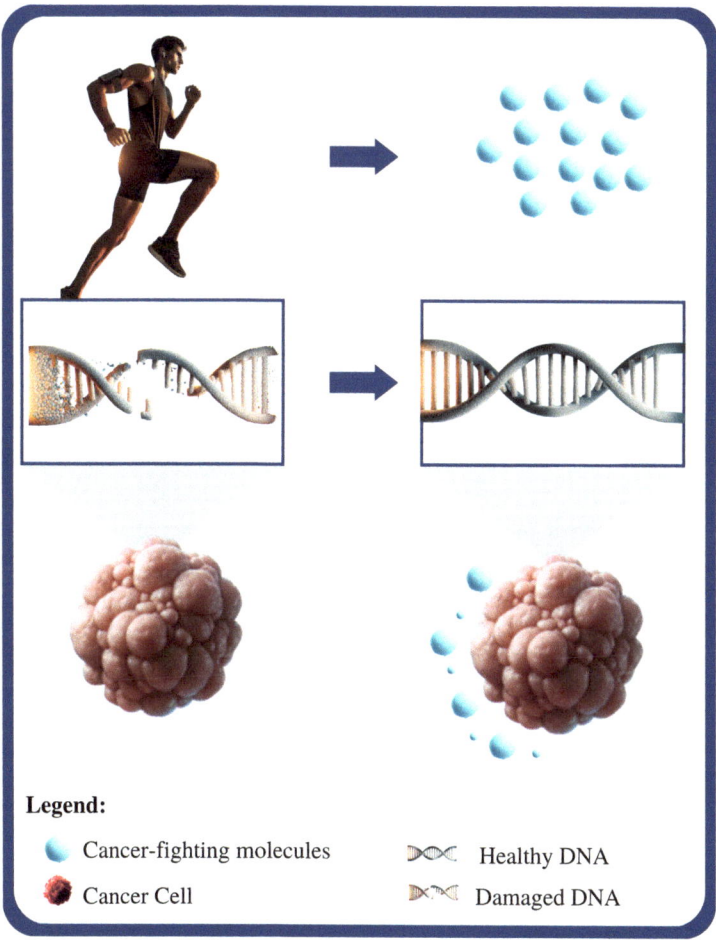

Legend:

Cancer-fighting molecules   Healthy DNA

Cancer Cell   Damaged DNA

Fig. 9.3 Exercise changes our body's chemical composition. Physical activity releases certain cancer-fighting chemicals that interact with abnormal cells, stimulate DNA repair, and decrease cell proliferation

9.2   Exercise and Side Effects

Physical activity improves physical and mental health but should be practiced in moderation, avoiding over-intensive sports that can lead to muscle strain and injury or exhaustion and fatigue. Physical activity releases oxygen free radicals into our system and can cause oxidative stress. This type of cellular damage can lead to mutations in the lipids, proteins, and DNA that make up our cells. It can even promote the proliferation of cancerous tumor cells. The good news is that our bodies have built-in defenses to combat the effects of these free radicals. Suppose we exercise only moderately and for a reasonable length of time. In that case, we can manage the influx of oxygen-free radicals without significantly damaging our cells and tissues.

Anyone can have a difficult time dealing with the physical and emotional effects of cancer treatment. Treatment is an extremely important event that can significantly impact one's quality of life [21]. An intensive treatment regime often gives us side effects such as general fatigue, lack of appetite, loss of muscle mass, and reduced capacity for physical activity. In fact, scientific studies indicate that roughly one-third of the problems associated with cancer stem from the patient's inability to maintain physical activity during treatment.

Many patients who survive cancer experience troubling side effects from their treatments. The side effects may include weight gain, endocrine system malfunction, and the risk of developing cardiovascular diseases. Thankfully, these burdens can be lessened by regular physical exercise. The primary goals at this point should be to regain self-confidence, reestablish physical performance and reinforce a positive outlook on life. Multiple epidemiological studies have indicated that engaging in regular physical activity improves self-reported health, reduces fatigue and shortness of breath, and facilitates better memory recall.

Although exercise is often the last thing on a patient's mind after they have had cancer treatment, analytical studies overwhelmingly demonstrate its positive effects. For example, 22 of 24 separate studies demonstrated that exercise could significantly help reduce posttreatment fatigue [22]. This trend was clear in elderly patients who had sports therapy while being treated for cancer. Indeed, professionally supervised sports therapy is becoming an increasingly important part of cancer treatment and rehabilitation. Many hospitals are now beginning to offer their patients tailored physiotherapy treatments to help them get exercise and manage their posttreatment fatigue.

9.3   Exercise and Physical Activity

Physical activity, such as exercise, improves our physical and mental health, but we must do it in moderation. Getting too much of it is like having a medical overdose, leading to more harm than good. Going overboard with intense activities can strain muscles and cause injuries. This is because we are overworking the muscles or ligaments. Doing too much too soon can also overwhelm our bodies, leading to burnout and fatigue. So, when it comes to exercise, intensity, type, and duration should all be considered, as we must practice moderation to reap the most benefits.

Carcinogenesis means cancer cell formation, and exercise is essential for keeping carcinogenesis at bay. Exercise helps balance the energy within our bodies. Our body considers how much energy we consume and how much we expend through activity. This is our energy balance. By increasing our physical activity and reducing how much food and calories we consume, we can significantly guard against cancer growth. Exercise also has other important health benefits, such as strengthening our muscles, boosting our mood, and improving our cardiovascular fitness [20]. Finally, physical activity has been identified as a key factor in cancer risk reduction thanks to its positive impact on obesity and body fat distribution. By engaging in physical activity or exercise, we can reduce our abdominal fat and overall obesity level, thereby decreasing the risk of getting certain types of cancer.

9.4   Cancer Variables

Cancer survivors' exercise levels may differ from the standards of the general population. Oncologists and general physicians might not give the same advice when it comes to physical activity and exercise. Therefore, it is essential to consider multiple variables when tailoring an exercise program that is right for that individual. Many variables will likely have an impact on the recommendations doctors make, such as age, gender, particular cancer type and stage, treatment received, and dose of treatment, among other factors. Due to this complexity, guidelines help doctors determine the best course of action for each survivor's unique needs. Taking on any form of physical activity is important for healthy living and managing the side effects of cancer treatment [23].

Many people find it difficult to achieve the goal of being physically fit. However, we can come close to this ideal by incorporating simple exercises into our daily routine, such as walking or cycling to work or setting up an exercise bike in front of the TV and pedaling away while watching our favorite series.

Recent large-scale studies have suggested that regular exercise for at least 3 h a week can benefit cancer patients significantly. ***Moderate-intensity activities*** might involve walking at a speed of 3 miles per hour (5 km/h) or biking at a speed of up to 20 miles per hour (32 km/h). ***Vigorous exercise*** would mean running 5 miles per hour (8 km/h) or faster, as well as biking faster than 20 miles per hour (over 32 km/h). Although light exercises, such as brisk walking, can be beneficial, vigorous exercises have been found to decrease mortality rates. Further research has indicated that just 30 min of physical activity a day can already improve physical performance and health noticeably. Currently, it is recommended to combine aerobic exercises, such as running or biking, with two sessions of resistance training and weightlifting for 150 min each week [23].

9.5   Sports Therapy: More Than Just a Chance

Sports offer us many ways to improve our health and well-being. Whether it's swimming, dancing, hiking, or a program with some training equipment—you can find the best way for you to get active. If you've never been into sports or physical activity before, this is your chance to start something new. Being confronted with illness is often an opportunity to start or resume physical activity, even for patients who have never done so before. Even if it's not always easy to motivate yourself at first. Sports therapy can be a source of motivation. The key is to identify personal goals so that exercise becomes enjoyable and successfully helps us reach our goals. Friends or family can also be included for some extra support. Additionally, even those who used to play sports before becoming unwell may still do so with adapted programs designed for specific needs. Regular exercise is crucial for maintaining good health. However, it's essential to remember that everyone has different physical abilities and schedules. Ideally, we recommend, as written abore, combining aerobic exercise, such as running or cycling, with two strength training sessions per week, totaling 150 min. We understand that this routine may not be achievable for everyone. However, it's crucial to get as close to this ideal as possible to reap the benefits of exercise. Every little bit counts, and even small changes in your routine can make a significant difference in your overall health and well-being. You can motivate yourself by identifying personal goals even if you have never been active. Achieving these goals can help boost your self-esteem and make you feel better physically and mentally. You could start by cycling in front of the TV daily to regain enough strength to run errands on your own or to go for walks with your pet or loved one. Going to the gym several times a week can also help you feel more comfortable in your body. Meeting up with your friends at the aqua gym or playing tennis with a buddy are other great ways to stay active. Losing a few kilos can also help reduce knee pain, and enjoying a nature bath while hiking is excellent for your overall health. Combining physical activity, friendship, and body conditioning can make exercise enjoyable and beneficial on multiple levels.

9.6   Setting Goals

Regular exercise can be a simple yet powerful way to gain autonomy and achieve long-term health goals. Begin with a 30-min walk every day, rain or shine, which can be divided into two 15-min sections. Stroll to a nearby tea-room, grab a coffee, and then head home. Or, walk to the park, take a 15-min breather on a bench, and continue on your way. If you're already engaged in physical activity, consider attending a training session and stopping as soon as you feel tired or challenged. The benefits of regular exercise cannot be overstated. It can improve your overall condition, health, self-confidence, and self-esteem and enhance your reaction time, balance, and independence in everyday tasks. For cancer patients and survivors, regular physiotherapy sessions can help achieve these goals by providing the necessary guidance to engage in physical activity and build confidence in their abilities. Don't let apprehension or fear hold you back. Take the first step toward a healthier, happier you today.

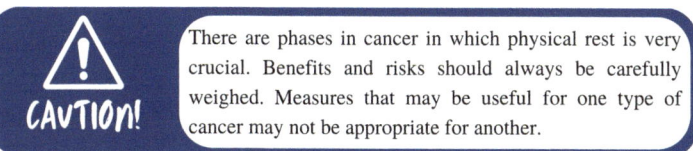

CAUTION! There are phases in cancer in which physical rest is very crucial. Benefits and risks should always be carefully weighed. Measures that may be useful for one type of cancer may not be appropriate for another.

9.7   The Right Type of Training

No matter what your ultimate goals may be, choosing the right type of training is key to achieving success. When deciding which activities are right for you, it is important to consider your individual capabilities, needs, and preferences. This includes deciding whether using aids such as machines or equipment will benefit your exercise routine or if it is better to attend a fitness studio or gym where all necessary resources can easily be found. Although you may think that you can figure it out on your own, we recommend you work together with a professional who can ensure that you maintain proper postures during exercise while also providing you with support and feedback to reach the optimal result.

Endurance Training

Endurance training is an effective way to improve both your physical and mental health. While similar to aerobic exercise, it takes a more sustainable approach for those looking for long-term gains. To do it correctly, aim for at least two to three weekly sessions consisting of medium-intensity activities lasting between 15 and 45 min. It is important to have a certain level of fitness to participate in such a program. However, not everyone is able to maintain the discipline required. If you persevere and stay committed, you will start noticing positive changes, such as improved stamina, in just a few weeks. For those who need to focus more on protective measures due to pre-existing medical conditions, there are alternatives like low-intensity daily training and interval sessions that alternate short bursts with rest phases.

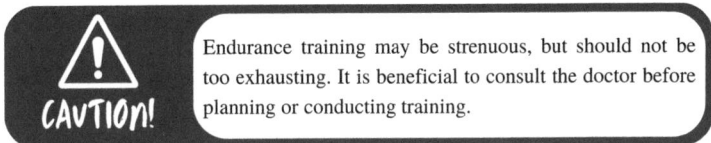

CAUTION! Endurance training may be strenuous, but should not be too exhausting. It is beneficial to consult the doctor before planning or conducting training.

Strength Training

After going through treatments for cancer, many people end up feeling weaker than before. It is normal for them to find themselves easily tired. Yet, this can then lead them to stop their regular activities, which can sometimes lead to a sense of helplessness. In turn, inactivity can cause muscles to become limp and the overall muscle mass to decrease. However, it's important to remember that just as the mind can impact the body, it can be challenging to engage in regular activity if you're not feeling motivated. Conversely, the mind is the key to performing at any level of sport. The body can also influence the mind: By feeling better about your body and realizing that you're gaining autonomy and independence through regular exercise, you can significantly boost your self-confidence.

Strength training makes you stronger and fitter. It is important to practice these exercises with care. If you have a risk of bone metastasis, osteoporosis, or fracture, you should not do intense strength training. After getting clearance from a health professional, make sure you choose exercises that are suitable for your abilities. Practice 2–3 times a week, doing 6–10 different exercises each time. Before each session, warm up properly to avoid injuries and ensure your breathing is in check—inhale when relaxing and exhale during exertion. Perform all reps slowly, without sudden movements, and use only half of your strength to feel a slight effort. You can adjust resistance settings on machines accordingly. All of this will help you get the most out of your strength training. Seek authorization and support from a health professional before embarking on a fitness program is recommended to ensure a safe and effective workout routine. By doing so, we can maximize the benefits of exercise and live healthier, happier lives. It's important to adjust the intensity and type of exercise to suit each individual's needs. If there is a risk of bone fractures due to conditions such as bone metastases or osteoporosis, it is recommended to seek authorization and guidance from a health professional before starting a fitness program. This can help prevent fractures in fragile bones.

9.8   Coordination, Mobility, and Relaxation

Mobility and coordination activities are just as essential to our overall health and well-being as endurance and strength training. Exercises such as stretching can make regular daily activities easier and improve the quality of our movements, which is also essential to maintaining agility, coordination, and suppleness. Additionally, it can be incredibly beneficial for your body when you incorporate short periods of absolute relaxation into your schedule—before or after exercise or instead of exercise altogether. These breaks help your body rest and recharge. Pleasant activities that yield a sense of full mental and physical recovery include taking warm baths, listening to your favorite music, breathing exercises, mental training, visualization, meditation, massages, or any activity customized specifically for your needs.

Many activities can provide gentle strengthening and coordination of individual muscle groups, such as qigong, tai chi, chi ball, yoga, pilates, circle dancing, autogenic training, and shiatsu. To receive maximum benefit from these exercises, it is important to focus on form and technique. It is also critical to ensure the activities are adapted to everyone's body type and fitness level. Deep breathing should also be maintained to integrate physical practice fully with the relaxation of the mind and body. Taking care of your mental and physical health is essential in today's fast-paced world. Engaging in activities such as hot baths, listening to music, breathing exercises, visualization, mindfulness meditation, and massage can help you unwind and feel rejuvenated. These sports and relaxation activities are suitable for your body, enhance your body awareness, and help you focus on the present moment. You can manage your anxiety and reduce stress by training yourself to be present. Engaging in physical activity and setting goals for yourself can give you a sense of accomplishment and control over your life. Many patients feel like they have lost control and fear what might happen. But taking control of your body and mind can help you feel empowered and confident. Whether you're looking to get out of the house or take on a sporting challenge, any activity that enables you to achieve your goals will help you feel more in control. So why not take the time to prioritize your mental and physical health? Engaging in physical activities, sports, and relaxation techniques can help increase body awareness and allow us to focus on the present moment. When we concentrate on physical activity, even if it is just paying attention to our breathing, we are less likely to worry about the uncertain future. This practice of being present can reduce our tendency to worry about what's to come and provide a sense of relief. Practicing physical activities and setting achievable goals can help us regain a sense of control, which can counterbalance the feeling of losing control that many patients experience when they are anxious about the future. Any activity that allows patients to set and achieve goals, whether it's leaving the house or accomplishing a sporting challenge, can help them regain a sense of control and improve their overall well-being.

The Language Test

If you're unable to speak clearly or comfortably during exercise, it could be a warning sign that the intensity level is too high. You can use the language test to determine how intense your exercise should be. The language test means being able to talk without feeling out of breath or having side stitches. When necessary, slow down your breathing and reduce the weight you are lifting, for example, to make sure you are training at the right level.

The Optimal Exercise Program

An exercise program tailored to your needs will help you improve your overall physical fitness. It will also increase your strength and endurance and help you become more mobile and relaxed. However, you need to keep in mind any personal limitations that may arise and stop a particular activity if it gives you any pain or discomfort. To continue training without overexerting the body, try decreasing the intensity of the exercises while increasing the duration at the same time. Additionally, remember to take breaks when you need them so your body can recover optimally. Staying hydrated throughout your workouts is also greatly recommended. You can do this by drinking tea, diluted juice, or preferably water. In general, taking part in a sports and exercise program at least three times a week for an hour each time can enable you to strengthen muscles and increase your endurance. It's not possible to do a 1-h session without feeling tired. It's much better to try training regularly, even for a short session. Even 10–15 min a day is helpful, rather than giving up by telling yourself you can't do it. Yes, even 10–15 min a day is helpful!

9.9 Cancer Sports Group

Joining a cancer sports group can be incredibly beneficial for individuals battling cancer and cancer survivors alike. It allows individuals to take control of their recovery process by setting their own goals for exercising. At the same time, participating in sports in groups gives people the chance to socialize and connect with others in similar situations. Mutual understanding is extremely powerful, as less explanation is needed when talking about shared experiences. This is one motivation for people to join a cancer sports group. Ultimately, it can also improve our adherence to the physical activity program and our quality of life [24].

Facilitators who have specific experience working with cancer sports groups assist in assessing what the participants are capable of. Facilitators do this to make sure that the program is adjusted to meet the participants' physical needs. This could involve improving coordination, balance, or mobility, as well as increasing their stamina and strength levels. Before taking part in any activity, it is always important to let the facilitator know if you experience any issues, such as low blood pressure or increased risk of fractures due to osteoporosis. Often, you will need permission from your doctor for certain activities, too. Other patients may prefer to join a "conventional" club to get out of the cancer environment and talk about something else. Many clubs have programs for beginners, enabling them to join. Finally, getting back to their sports club, where they've already made friends, will be the priority for other patients. A medical certificate is sometimes needed for sports and competitions.

Cancer patients and survivors often desire to take up sports as part of their physical recovery plan. However, sometimes they may be hesitant about joining a team because of negative emotions, such as low self-esteem, body image issues, and constant intrusive thoughts about their illness. In such cases, a professional trainer can help them by providing support, social interaction, and motivation, which are crucial for successful rehabilitation. Talking to a professional whenever you have questions is always a good idea. This is true when joining a cancer sports group, too.

9.10 Exercises to Do at Home

These days, working out doesn't require leaving the safety of your home. At home, you can still take part in activities that boost your well-being and lift your energy level. Simple exercises are great—like stretching to keep your spine and upper body nimble and strengthening your back, legs, shoulders, bottom, and abs. Therefore, it's easy at home to practice movements that promote shoulder mobility while focusing on your breath, alternating between tensing up and letting go. These highlights of basic physical fitness help us stay strong, sharp, and gorgeous. Once again, while exercising, make sure to pay attention to your breathing. Stop immediately if you feel any pain. Allow yourself short breaks while keeping up a steady rhythm throughout the training session. Many simple exercises can be continued at home (between physiotherapy sessions or afterward) to increase energy levels, mobility, strength, and balance.

9.11 Daily Life as Exercise

Increasing your physical activity can be easy. It does not mean only having to make an arduous undertaking by attending classes or joining sports groups. You can easily inject exercise into your daily life with simple and fun activities that you will enjoy. For example, try taking the stairs as opposed to using the elevator. Try taking walks around your neighborhood instead of driving or riding to nearby places. Try going for a stroll in nature with family and friends.

Or you can even try playing ball games with the kids. These activities will help speed up your journey to fitness and ignite a passion and enthusiasm within you that goes beyond simply losing weight. Whether it is dancing, running, or moving around more—make exercise an exciting part of your life by exploring new avenues of physical activity that thrill you. Climbing stairs helps you keep your muscles strong. As discussed above, particular attention should be given to breathing, which must not be interrupted, and speech should be allowed.

9.12 A Starting Guide for Patients

Becoming physically active is a great way for you to improve your physical and mental health. But the journey of getting there doesn't have to be overwhelming. A great place to start is by honestly assessing yourself and working out how much time you can dedicate to your physical activity each day or each week. You can begin with simple steps such as walking more during the day, stretching or doing physiotherapy exercises at home, or participating in activities such as yoga that can help gradually increase your fitness level. Even if you no longer have the same energy or flexibility, taking inspiration from physical activities you practiced in youth can be helpful. Talking about it with friends and family and joining a class or action together (why not join a gym class, table tennis club, or go swimming together)? Once you know what type of physical activity you like and how much time you can spend on it, then it's all about finding ways to stay motivated and stick with your plan. And don't forget that it's okay if your progress is slow—every small step counts! Regular rewards help keep people motivated. Everyone needs to find their prize. If I reach my goal, I can also enjoy an outing. Remember, even if things aren't moving quickly, every little step counts!

9.13 The Exercise Diary

Creating a training diary can be an extremely beneficial tool for recording your physical activity as part of your cancer sports group. Keeping detailed records such as feelings, achievements, difficulties, and activity intensity will help you listen to your body and decide which activities you enjoy the most. Asking the cancer sports group for a template of this kind of record is a great first step. It helps you get started with a personalized diary that reflects your individual journey. This record will ultimately give you more control over how different exercises impact your well-being. It will also help guide you in constructive conversations you can have with caretakers and healthcare providers.

9.14 Improving the Quality of Life

Regular exercise should not be overlooked. It improves our overall sense of well-being and increases our mobility, strength, and endurance. Regular exercise can also help bolster our body's natural defense mechanisms. It may also minimize the residual impacts of cancer treatment, such as fatigue. This, in turn, enhances our performance and confidence and thus greatly improves our quality of life. In sum, regular physical activity clearly fosters better mental health and physiological wellness [12].

Both aerobic and resistance training are effective because they can slow down cancer cell growth and improve physical fitness. Exercise also boosts the immune system and enhances our mood and overall quality of life. It's recommended that you create a personalized exercise program that includes aerobic, strength, and flexibility exercises. Joining cancer sports groups can provide additional support and motivation. Finally, the chapter emphasizes that any form of exercise is beneficial and there are simple ways to make daily life more active.

References

1. Schrack JA et al Understanding physical activity in cancer patients and survivors: new methodology, new challenges, and new opportunities. Cold Spring Harb Mol Case Stud 2017;3(4).doi:https://doi.org/10.1101/mcs.a001933
2. Global cancer facts & figures. https://www.cancer.org/research/cancer-facts-statistics/global-cancer-facts-and-figures.html. Accessed 05 July 2024.
3. Wiseman M et al.Food, nutrition, physical activity and the prevention of cancer: a global perspective. Summary. 2007.
4. Pedersen BK, Saltin B. Exercise as medicine—evidence for prescribing exercise as therapy in 26 different chronic diseases. Scand J Med Sci Sports. 2015;25(Suppl 3):1–72. https://doi.org/10.1111/sms.12581.
5. Folkman J, Kalluri R. Cancer without disease. Nature. 2004;427:787. https://doi.org/10.1038/427787a.
6. McTiernan A, et al. Physical activity in cancer prevention and survival: a systematic review. Med Sci Sports Exerc. 2019;51:1252–61.

7. Miyamoto T, et al. Effect of post-diagnosis physical activity on breast cancer recurrence: a systematic review and meta-analysis. Curr Oncol Rep. 2022; https://doi.org/10.1007/s11912-022-01287-z.

8. Spei M-E, et al. Physical activity in breast cancer survivors: a systematic review and meta-analysis on overall and breast cancer survival. Breast. 2019;44:144–52. https://doi.org/10.1016/j.breast.2019.02.001.

9. Friedenreich CM, et al. Physical activity and survival after prostate cancer. Eur Urol. 2016;70(4):576–85. https://doi.org/10.1016/j.eururo.2015.12.032.

10. Rocha-Rodrigues S, et al. Skeletal muscle–adipose tissue–tumor axis: molecular mechanisms linking exercise training in prostate cancer. Int J Mol Sci. 2021;22(9) https://doi.org/10.3390/ijms22094469.

11. Matta K, et al. Healthy lifestyle change and all-cause and cancer mortality in the European prospective investigation into cancer and nutrition cohort. BMC Med. 2024;22(1):210.

12. Kesting S, et al. Exercise as a potential intervention to modulate cancer outcomes in children and adults? Front Oncol. 2020;10:196. https://doi.org/10.3389/fonc.2020.00196.

13. Cole SW, Sood AK. Molecular pathways: beta-adrenergic Signaling in cancer. Clin Cancer Res. 2012;18(5):1201–6. https://doi.org/10.1158/1078-0432.CCR-11-0641.

14. Elenkov IJ. Glucocorticoids and the Th1/Th2 balance. Ann N Y Acad Sci. 2004;1024(1):138–46. https://doi.org/10.1196/annals.1321.010.

15. Run for your life: exercise protects against cancer. ScienceDaily. https://www.sciencedaily.com/releases/2016/04/160407121459.htm. Accessed 04 July 2024.

16. Hojman P, et al. Exercise-induced muscle-derived cytokines inhibit mammary cancer cell growth. Am J Physiol Endocrinol Metab. 2011;301(3) https://doi.org/10.1152/ajpendo.00520.2010.

17. Roy P, et al. Exercise-induced myokines as emerging therapeutic agents in colorectal cancer prevention and treatment. Future Oncol. 2018;14(4):309–12. https://doi.org/10.2217/fon-2017-0555.

18. Vu TT, et al. The role of decorin in cardiovascular diseases: more than just a decoration. Free Radic Res. 2018;52(11–12):1210–9. https://doi.org/10.1080/10715762.2018.1516285.

19. Manole E, et al. Myokines as possible therapeutic targets in cancer cachexia. J Immunol Res. 2018;2018:8260742. https://doi.org/10.1155/2018/8260742.

20. Westerlind KC. Physical activity and cancer prevention--mechanisms. Med Sci Sports Exerc. 2003;35(11):1834–40. https://doi.org/10.1249/01.Mss.0000093619.37805.B7.

21. Courneya KS. Physical activity and cancer survivorship: a simple framework for a complex field. Exerc Sport Sci Rev. 2014;42(3):102–9. https://doi.org/10.1249/JES.0000000000000011.

22. Cormie P, et al. The impact of exercise on cancer mortality, recurrence, and treatment-related adverse effects. Epidemiol Rev. 2017;39(1):71–92. https://doi.org/10.1093/epirev/mxx007.
23. Lemanne D, et al. The role of physical activity in cancer prevention, treatment, recovery, and survivorship. Oncol Williston Park. 2013;27(6):580–5.
24. Abdin S, et al. A systematic review of the effectiveness of physical activity interventions in adults with breast cancer by physical activity type and mode of participation. Psychooncology. 2019;28(7):1381–93. https://doi.org/10.1002/pon.5101.

Epilogue

This book has taken us into the breathtaking world of cells, which, despite their small size, have an unimaginable complexity and form the foundation of life. We have marveled at the astonishing diversity of cell functions and the precision of the cell cycle that forms tissues and organs in the human body. At the same time, we have realized that disruptions in this cycle can lead to uncontrolled cell growth and cancer. We have gained profound insights into the nature of cancer, its causes, and its characteristics. The critical role of DNA mutations, especially driver mutations, has been highlighted, as well as the influences of the microenvironment and inflammation on carcinogenesis. We have learned that a variety of risk factors, from genetic predispositions to environmental influences, contribute to the development of cancer. Cancer prevention has proven to be an essential component in the early detection and prevention of cancer. Screening procedures such as mammograms, Pap tests, and colonoscopies are crucial as they can detect precancerous conditions and early stages of cancer. Regular use of these tests can save and prolong lives. The emotional and psychological burden of a cancer diagnosis has been exposed. We have come to understand that while cancer is no longer necessarily a death sentence, it can still cause anxiety, stress, and depression. The strength of family, friends, medical staff, and support groups is essential in this fight. After diagnosis, doctors work on tumor boards to determine the optimal treatment. Each therapy must be individualized, as cancer and patient response vary, providing reassurance and confidence. Local treatments such as surgery and radiotherapy are successful for some patients, while others benefit from chemotherapy. Cancer survivors face the challenge of coping with the consequences of survivorship and possible late effects of treatment. Survivorship plans and regular follow-up care are essential to detect and treat relapses and secondary diseases at an early stage. The importance of a healthy diet for cancer prevention and treatment was emphasized. A diet rich in fruit and vegetables can reduce the risk of cancer, while processed meat and alcohol increase it. Physical activity has been shown to be a powerful ally in cancer prevention and treatment. It can reduce the risk of cancer, improve survival rates, and alleviate side effects of treatment. In conclusion, we recognize that the fight against cancer is a multifaceted approach that includes prevention, early detection, tailored therapies, psychological support, and lifestyle changes. Each step along the way is significant and leads us to a healthier future where cancer is no longer seen as an invincible enemy but as a challenge that we can face together with courage and determination.

10.1.1 Get in Touch

CANCER_SIMPLY_EXPLAINED

Index